THE GARDENER'S CATALOGUE
SEX in the Garden

THE GARDENER'S CATALOGUE

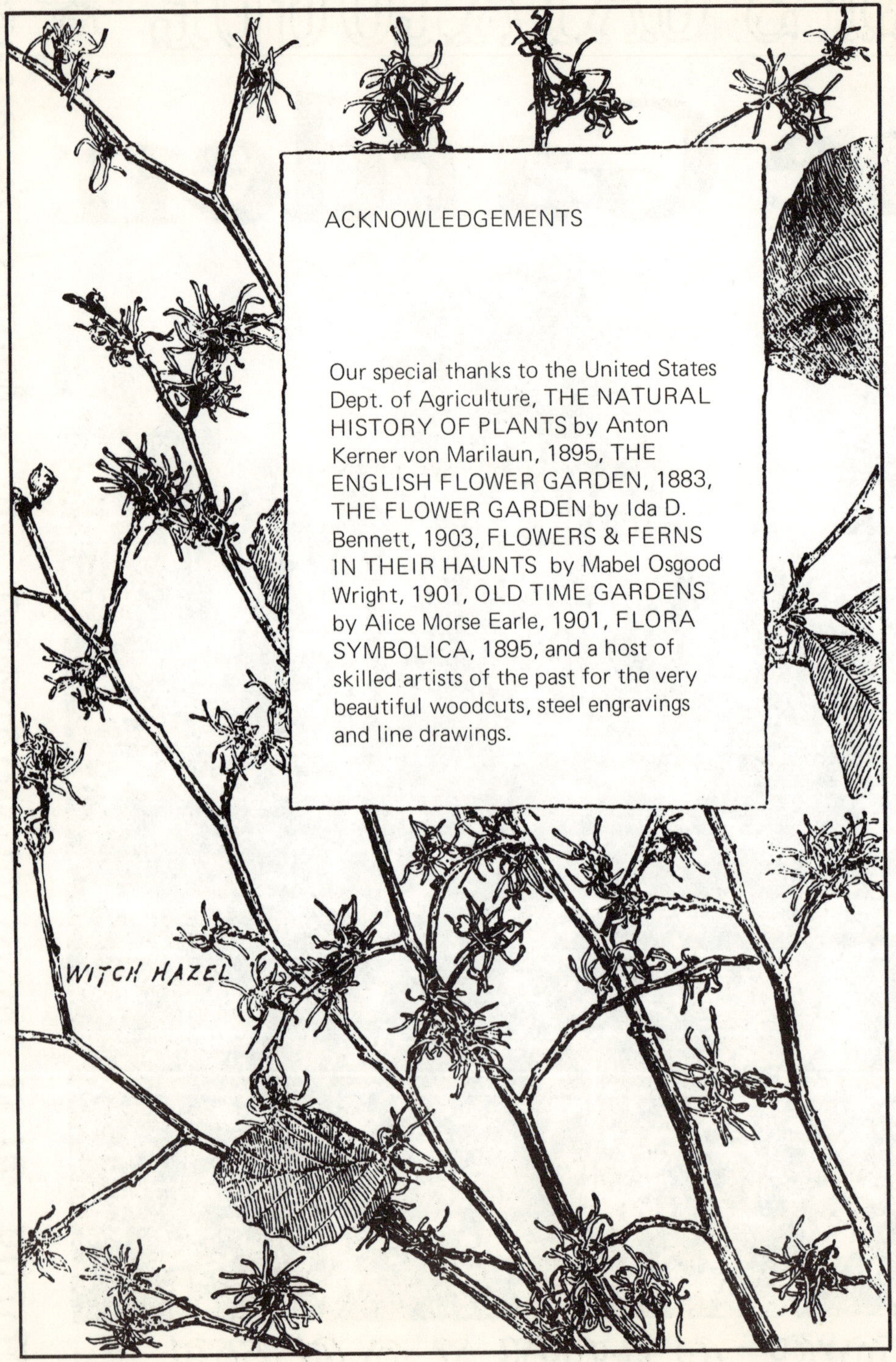

COPYRIGHT © 1976 by Tom Riker and Harvey Rottenberg

ACKNOWLEDGEMENTS

Our special thanks to the United States Dept. of Agriculture, THE NATURAL HISTORY OF PLANTS by Anton Kerner von Marilaun, 1895, THE ENGLISH FLOWER GARDEN, 1883, THE FLOWER GARDEN by Ida D. Bennett, 1903, FLOWERS & FERNS IN THEIR HAUNTS by Mabel Osgood Wright, 1901, OLD TIME GARDENS by Alice Morse Earle, 1901, FLORA SYMBOLICA, 1895, and a host of skilled artists of the past for the very beautiful woodcuts, steel engravings and line drawings.

WITCH HAZEL

All rights reserved. No part of this book may be reproduced or utilized in any form or by any means, electronic or mechanical, including photocopying, recording or by any information storage and retrieval system, without permission in writing from the Publisher. Inquiries should be addressed to William Morrow and Company, Inc., 105 Madison Ave., New York, N. Y. 10016.

Printed in the United States of America.

Sex in the Garden

The GARDENER'S CATALOGUE series was conceived and is independently produced by Harvey Rottenberg and Tom Riker.

SEX IN THE GARDEN was edited and designed by Tom Riker, whose years of experience as a horticulturist and botanical art specialist made the book possible.

John Krausz was production man, and did the typography and photoconversions. Special thanks to Lars Skattebol for his editorial and technical knowhow; Lenore Stein, actor and horticultural researcher; and Michael Edstrom, our board person.

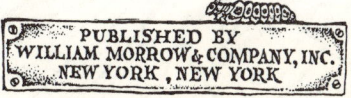

PUBLISHED BY
WILLIAM MORROW & COMPANY, INC.
NEW YORK, NEW YORK

TABLE OF CONTENTS

INTRODUCTION
THEORY AND PRACTICE 1-17

Description & Classification of Plants 9-11, Linnean System 12-13, Object of Botanical Research 15-17

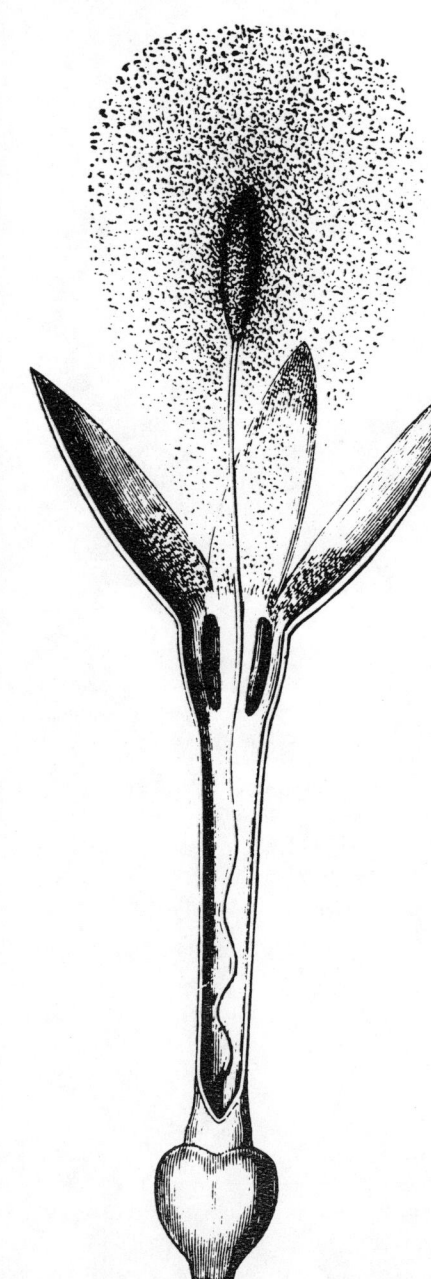
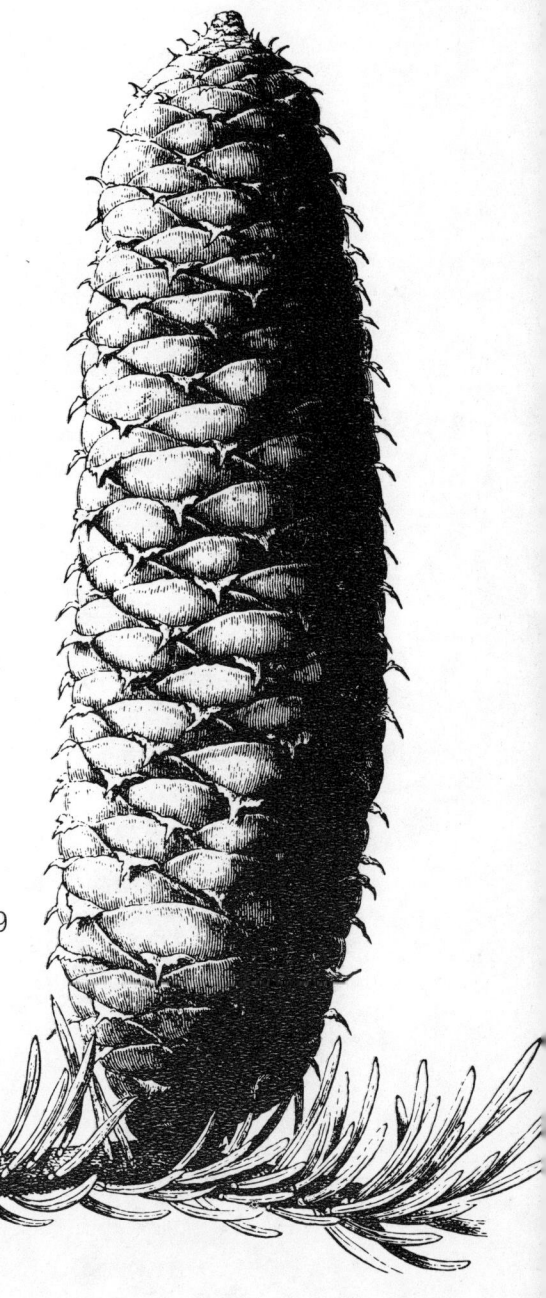

PROPAGATION 18-51

Propagation by Seed 19-21, Propagation by Bulbs, Corms & Tubers 22-29 Propagation by Layers 31, Home Propagation of Trees & Shrubs 32, Air Layering 33, Layering the Verbena and Carnation 34, How to Dwarf a Dracaena 35, Cuttings 37-41, Propagation by Division and Runners 42, Propagation by Roots & Leaves 43, Grafting & Budding 44-46, Inarching 47-48, How to Trim a Coleus 49, Transplanting 50-51

PLANT SEXUAL SYSTEMS 52-111

Introduction 53, Analytical Arrangement of Botanical Terms 55-85, Pollen & Seed 86-88, Bees, Butterflies & Blossoms: Our Useful Garden Insects 89-92, Protection of Pollen & Flowers 94-98, Interaction in Nature 100-108, Concealment of Honey 111, Ferns & Foliage 113-119

MEN & WOMEN IN THE GARDEN 120-159

Lead Figures 122-123, Knots & Parterres 124-129, Mazes 130-133, Topiary Work 134-137, Fountains 138-141, Sundials 142-143, Garden Leadwork 144-145, Terraces 146-149, Garden Houses 150-153, Gatepiers 154-155, Benches, Bridges, Shady Spots & Wells 156-159

FLORA SYMBOLICA 160-184

GLOSSARY 185-192

INTRODUCTION

THEORY AND PRACTICE

The overall idea of this book is the relationships between the animal world (both man and our four-legged friends) and the vast plant world. There is a theory and a practice in this plant world. Theory is the study of botany and the sub-sciences that make up the vegetable kingdom. Practice is horticulture and floriculture—the art of gardening, which as a hobby and activity is without a peer in most places of the world.

Most gardening books deal with the pragmatic values of gardening, never seeming to touch on the inner activity of the garden, such as perfume, color structure, and interaction of the animal world with the plant world. Yet the chemistry of the garden is a very balanced and delicate area where there is continual change both seen by the naked eye and microscopically.

The average gardener might ask why should he or she understand the interaction and unseen activity if the pragmatic result at the end of the season is a combination of flowers, vegetables and fruit. The rationale for such an understanding would be a desire for the knowledge of the chemistry of the garden in totality. This knowledge, in turn, should embrace the relationships of people and plants as well as the natural relationships that continue without the intervention of man.

I have tried to present in picture form some of the more interesting aspects of inner garden activity without getting into a heavy text of botany. The illustrations will cover only a fraction of the kinds of activity that take place twenty-four hours a day. But an understanding of this fraction brings more enjoyment of the overall scenic beauty. And more pleasure will be derived from the long hours one must spend in the garden to achieve the perfect rose or blue-ribbon melon.

There is little doubt that most people look upon the plant and animal world as game to shoot or to just observe. Trees are timber or shade and maybe firewood if you are lucky enough to have a fireplace. Vegetables are to be eaten, as are herbs and fruit. Farmers look at the land with a sense of provender for the stock, and flowers are to be cut for the table or to be photographed for the spring issue of your local gardening magazine. I have no basic quarrel with any activity as long as it does not harm the people or the ecology, but there are additional ways to see and appreciate our surroundings.

The general motivation for the study of plant life has been the need of the individual to satisfy his own desires, whether they be for food, beauty, or study as an intellectual activity.

The first types of classification were to determine whether given plants were nutritious or poisonous. There naturally followed the first attempts at cultivation and observation of various types of plant life.

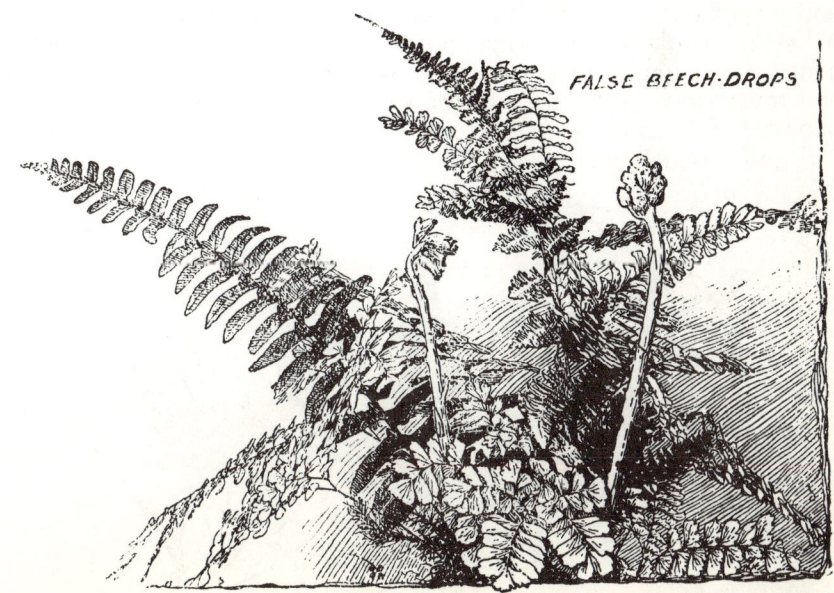

Next there came another powerful incentive to the study of herbs, roots, and seeds, and to the careful comparison of their differences. This incentive was the hope and belief that higher powers had endowed particular plants with healing properties. In classical Greece there was a special guild that dealt with plants, the Rhizotomoi. The members collected and prepared such roots and herbs as were considered to be curative. These labors were augmented by those of Greek, Roman and Arabic physicians, as well as of gardeners, vine-growers and farmers. Eventually there was a large mass of information about the plant world, and for a long time this stood as botanical science. In fact, as late as the sixteenth century plants were looked upon purely from the utilitarian point of view, and this was true not only of the people in general but also of professed scholars. In the books of that period we find a concentration on medicinal properties. Just as men lived in the belief that human destinies depended upon the stars, so flourished the notion that everything upon the earth was created for the sake of mankind. In particular, there was the belief that in every plant were forces which if liberated could be used to help or hurt man. People imagined that they could observe magical qualities in plants. It was believed that specific plants could be used to aid or cure specific parts of the body. The similarity in shape between a particular leaf and the human liver led to the belief that this leaf was capable of aiding in the cure of hepatic disease. A blossom found to be heart shaped was assumed to be capable of curing cardiac complaints. From this there arose what came to be known as the doctrine of Signatures. This was brought to its highest development by the Swiss alchemist Bombastus Paracelsus (1493-1541). The doctrine of Signatures played a big part in the sixteenth and seventeenth centuries in the belief in nostrums, and of course the belief, or mania, of nostrums continues to the present time. There is a strong inclination now, as in the past, in favor of the supernatural and mysterious rather than in the simple and natural. Actually, most people today, as in the past, consider botany as subservient to medicine and agriculture.

But in addition to botanical knowledge derived from necessities of life there is a second avenue to be explored because of man's sense of beauty. At first this sense had to be satisfied by the cutting and arrangement of wild flowers and foliage. But ultimately there was an interest in the cultivation of the more showy and ornamental of the plants, and in different countries and in different times the cultivation of beautiful plants corresponded to the prevailing standards of the beautiful.

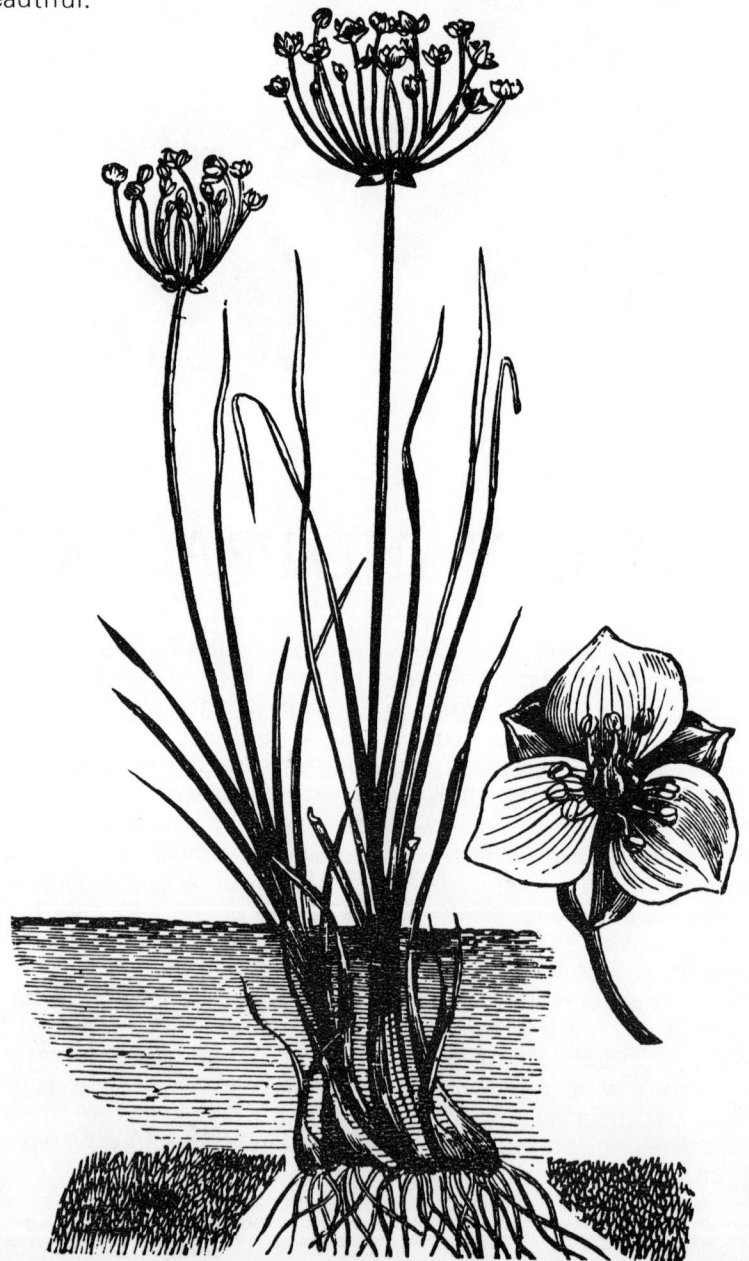

Arrow-head (*Sagittaria sagittifolia*).

THE DESCRIPTION AND CLASSIFICATION OF PLANTS

There is a third path leading to botanical knowledge. This springs from the impulse of those people who are endowed with a keen perception of form and want to investigate structural differences down to their tiniest characteristics. Workers in this field arrange and classify all distinct forms according to external resemblances. Names are given which seem to be appropriate to the position and importance of the gathered material. Catalogues are made and kept up to date.

This tendency of the human mind to gather and catalogue in relation to form has played a very important part in the history of botany. The first traces of this can be placed in a period long preceding our own scientific era, for there are descriptions and notes on comparative forms in the NATURAL HISTORY OF PLANTS, written by Theophrastus about the year 300 B.C. The material in the book is based on the observations of the Rhizotomoi and physicians and agriculturists. It is clear from the text of the book that in some cases the observers did seek out plants and try to arrange and catalogue them for their own sakes, the sakes of the patterns themselves, and not just for the economic and medicinal values of the plants.

During the Roman Empire, and the dark and middle ages, no one troubled himself about plants other than those known to be in some way useful. But when the countries of the west began to discover and treasure the findings of ancient Greek thought, there was a corresponding rise of interest in the practice of hunting for plants for the purpose of describing and cataloguing all distinguishable forms. Science had a renaissance at the same time as did art. Starting in the fifteenth century the plant enthusiasts were stimulated by the Greek writings on natural history. And while the ancient writings could not satisfy the thirst for knowledge, the direction of new discoveries took its course from what was enduring in the work of the Greeks—direct investigation of nature.

Stimulated by the old Greek writings on plants, searchers in both Northern and Southern Europe set about listing and grouping the different kinds of local plants. The results are preserved in a number of bulky herbal books. These folios date for the most part from the first half of the sixteenth century. The plants are described and discussed just as the authors happened to come across them. Only here and there can we find a feeble attempt to range together and make groups of nearly-allied species. Only passing attention was paid to the facts of geographical distribution. All jumbled together in a confused medley were plants native to the soil, herbs which flowered in gardens from seed bought from traveling vendors, and plants brought to Europe by explorers of the new western world.

Travel being slow and difficult, most researchers in botany were limited to their immediate surroundings. The researchers had only dim ideas of the extent to which floras of various places differed from one another. Centuries before, plants of the Mediterranean regions had been described by Theophrastus and Dioscorides and Pliny. The European researchers of the fifteenth and sixteenth centuries assumed that the plants in their areas must be the same as those mentioned by the ancients, even those plants, in say, Germany. The German "Fathers of Botany" (Brunfels, born about 1495, died 1534; Bock, 1498-1554; Fuchs, 1501-1566, are the best known) simply applied the old Latin and Greek names to species of plants growing in German areas. At first, these men were firmly convinced of the identity of German, Greek and Italian floras, and for awhile they persisted in this notion in spite of inconsistencies which were numerous and growing more so as researches progressed. It was only little by little that the botanists began to abandon their fruitless debates about Latin and Greek names for such plants as grew in northern Europe. Slowly it began to be seen that much as they were indebted for inspiration to the ancient naturalists, it was more important to look with a fresh and unbiased view at the greenery all around. And eventually, with the Germans well up in the lead, the botanists began to devote themselves to direct researches into their surrounding, native floras. Take for example the herbal of Hieronymus Bock, which appeared in 1546. He says in it, "the herbs growing in German countries are described from long and sure experience." Bock got himself involved in a big botanic controversy of the period, whether the Latin name Erica was applicable to the German heath. On this Bock said, "Be our heath the same as Erica or not, it is in any case a pretty and sturdy little shrub, beset with numerous brown rounded branches, which are clothed all over with small green leaves; and its appearance is like that of the sweet-smelling Lavender Cotton." At another place in the book Bock seemed to lose patience with those who concentrated too much on the names of things, and said that the proper thing to do would be to lay aside all disputes over nomenclature.

Greater Marsh Horse-tail

Common Basil.

A Belgian, Charles de L'Ecluse (1526-1609) was able to liberate himself entirely from the name-fighting of his times and the preceding ones of Bock. In his big work, which appeared at the end of the 16th century (with his name latinized to Clusius), he was guided mostly by a desire to become acquainted with every flowering thing, and not by a desire to quarrel over what each plant was called. He did his best to describe, and sometimes to draw, the various forms of plants. He also cultivated them and with some went to the length of preserving them in dried condition. Collections of dried specimens began to become popular in the time of Clusius, and such a collection first was called a "hortus siccus" and later on a "herbarium." And from then forward, all museums of natural history were furnished with them. Clusius was so possessed by the desire to see vegetation with his own eyes that he became the first botanist to do any considerable amount of traveling for the purpose of botanizing. He covered Europe all the way from Spain to the edge of Hungary and from the sea to the highlands of Austria. Other botanists also began to travel and eventually there were collections which contained specimens from much of the globe.

In this way an immense amount of material was gathered, and numerous isolated commentaries were written. Material was ready for a master botanist and classifier, and he appeared as the Swedish naturalist Linnaeus (1707-1778). He worked with enormous industry on his own researches and in checking those of others, and when he made surveys of the accumulated material he was able to gain almost universal recognition. Linnaeus introduced short names for the various species in place of the old cumbersome, awkward designations. He also refined earlier descriptions with brief, concise ones of his own. And Linnaeus marked out the different parts of a plant, such as root, stem, leaf, bract, calyx, corolla, stamens, pistil, fruit, and seeds. Refining still further he distinguished particular forms of these organs as, for instance, filaments, anthers, and pollen in the stamens, and also ovary, style, and stigma in the pistil. To each of these refined objects he assigned a technical name or terminus. Now, with the help of the terminology and descriptions Linnaeus pioneered in organizing, it became possible to recognize species from written descriptions, and to determine what names had been given the species by botanists, and to what group the species belonged.

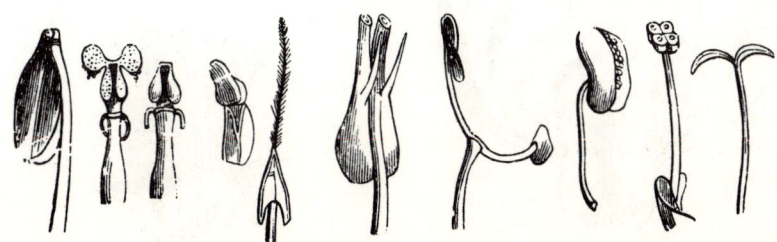

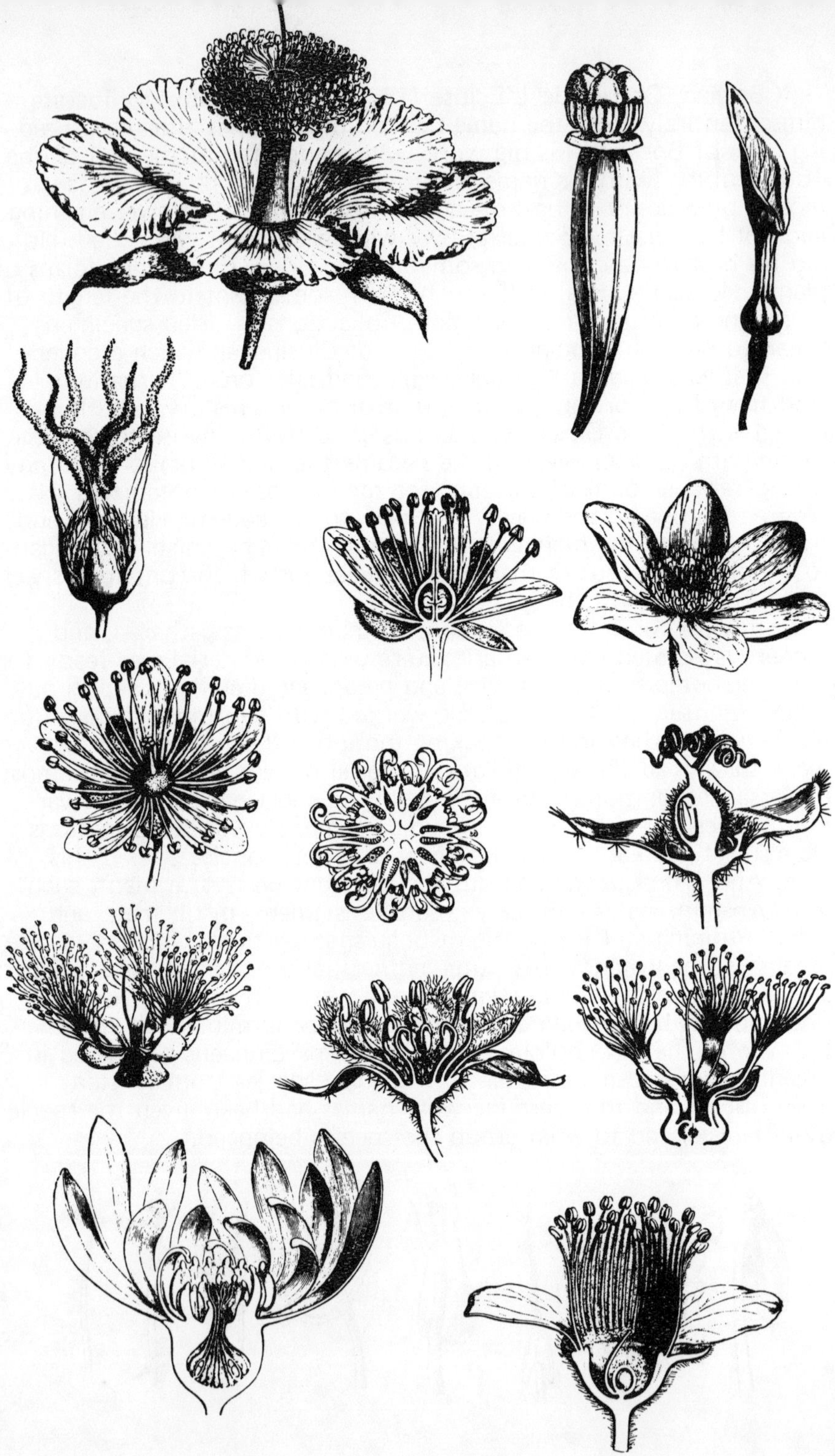

THE LINNEAN SYSTEM.

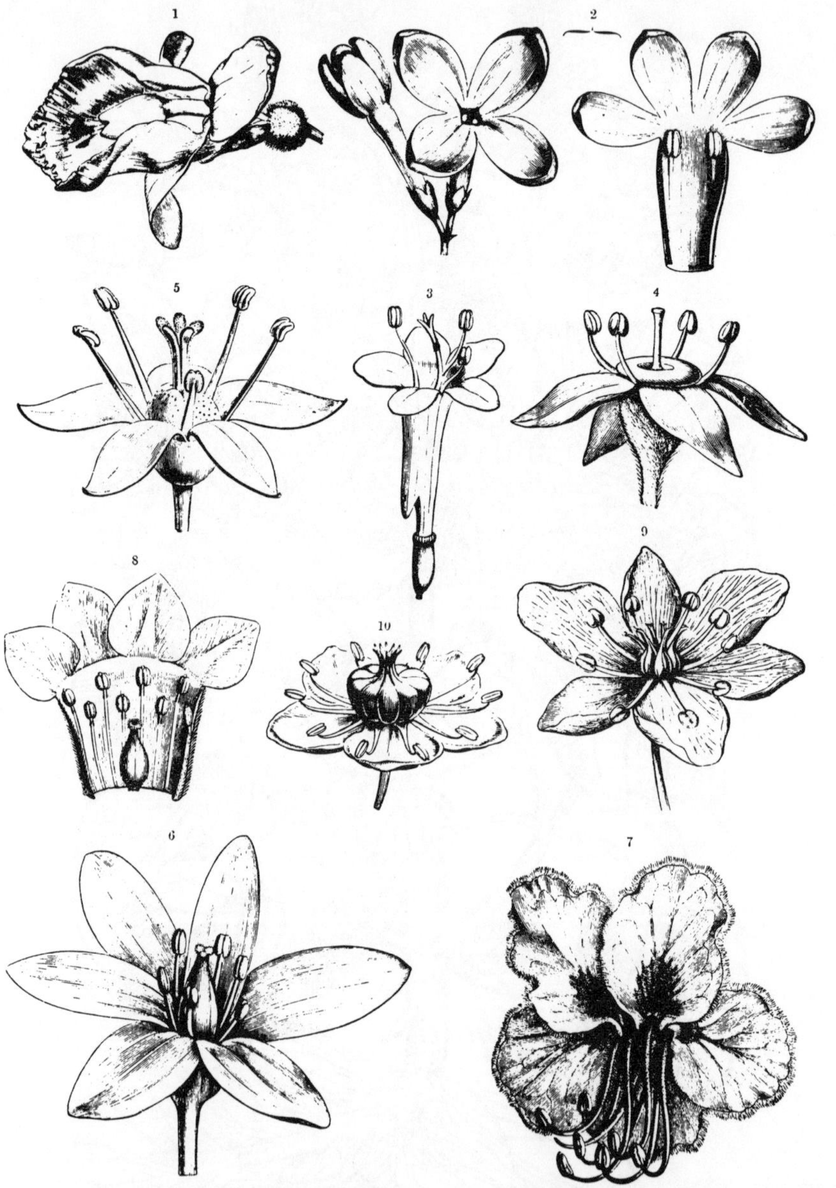

—Types of the 1st to 10th classes of the Linnean System.

¹ *Alpinia.*   ² *Syringa vulgaris.*   ³ *Valeriana officinalis.*   ⁴ *Cornus mas.*   ⁵ *Aralia Japonica.*   ⁶ *Gagea lutea.*   ⁷ *Æsculus Hippocastanum.*   ⁸ *Daphne Mezereum.*   ⁹ *Butomus umbellatus.*   ¹⁰ *Phytolacca decandra.*   All the flowers somewhat enlarged.

And Linnaeus, in the forefront as usual, finally was ready with his System. In it were established the characteristics of the various parts of a flower. In the System the number, relative length, cohesion, and disposition of the stamens formed the ground for division into Classes. Then within each Class were Orders, which were differentiated according to the nature of the pistil. Then with each Order came subdivisions into more closely defined groups called Genera. Linnaeus listed 23 Classes of Flowering Plants (Phanerogamia) and he added as a 24th Class one of Flowerless Plants (Cryptogamia). This 24th class was divided in turn into several groups (Ferns, Mosses, Algae, and Fungi) according to their appearance and the way they appeared.

Despite the slowness of communication of the 18th century, as compared with our own century, the System of Linnaeus took hold of botany all over Europe with amazing speed, and spread also to the civilized world in areas beyond Europe. Not just academics, but also educated laymen took to studying the botany of Linnaeus. Educated women did also. In France, Rousseau gave lectures on botany to a group of educated women. Goethe found botany had a strong attraction for him and it was called by many "the loveliest of sciences."

Linnaeus introduced the name "flora" to mean a catalogue of plants of a more or less definite district. Linnaeus himself wrote a flora of Lapland and Sweden, and this stimulated other botanists to do floras of their own areas. So by the end of the 18th century there were available floras of England, Piedmont, Carniola, Austria, and so on.

By following the system of Linnaeus, the botanists reached a certain peak of semi-perfection. Later on, unfortunately, many botanists tended to lose themselves in dull systematizing. Some would get involved in controversies over whether a plant described as a species should instead be called a variety. But the aberrations of this kind did not constitute a barrier to progress in botany. The passion of the botanists was such that traveling botanists ranged the world without any prospect of real material success from their discoveries. Those people not under the grip of the passion for botanizing cannot understand how anyone can devote half a lifetime to a classification of Algae or Lichens, or on a monograph about orchids. But enough enthusiasts became active to increase enormously the discovered and catalogued number of species. Theophrastus, about 300 B.C. mentions some 500 species, and Pliny about 78 A.D. had increased the number to 1,000. By the time of Linnaeus some 10,000 species of plants were known, and by the end of the 19th century the number approached 200,000. It should be emphasized that about half the species numbered since Linnaeus have been in the category of Cryptogams, that is, non-flowering plants. And the use of the microscope in modern times has done much to spread the discovery and systematizing of these non-flowering plants.

The microscope also did wonders in the discovery and study of flowering plants, and on the internal architecture of all species of flora. The intensive use of the microscope began with the opening of the 19th century, and this study sometimes was called the "inward construction of plants." It was the job of the plant anatomist to dissect plants, to look at the dissections under the microscope, and then to describe the various component parts as well as the general ground plan of what was being examined.

Blue Spruce

OBJECTS OF BOTANICAL RESEARCH
AT THE PRESENT TIME

Descriptive botany is concerned only with the configuration of a plant. What is called comparative morphology is the effort to trace back to a single prototype from the various forms exhibited by mature plants. And a third area for study is the history of development—the history of growth and differentiation of various forms. But these three paths of research and study leave relatively untouched the biological significance of the different forms. The hows and whys of the life processes themselves had to wait on developments in parallel sciences—physics and chemistry, mainly. And those hows and whys also had to wait on the time that botanists became convinced that basic understandings had to be disclosed by experiment and generalizations founded on many experiments.

Originally, botanical studies took the whole plant as first as an object of study, and next, its several parts. Lastly came the studies of cells and protoplasm. In modern research the direction of research is, in a way, completely reversed. First of all the histories of the ultimate organs are studied, then the significance of the various forms, and lastly what happens through the interplay of various kinds of animals and plants.

It is true that early attempts to define the biological, the final, significance of plants can take one back as far as Aristotle and his school. But the ideas of vegetable life at that time seem to us like fantastic dreams. The first actual experimental investigations into the vital forces of plants did not come until 1718 at the hands of Stephen Hales. And it was a full hundred years later before this kind of experimental research became comparatively common. In the 18th century, at last, there came the conception of a cell in plants as a miniature laboratory, a chemical laboratory. And research concentrated on attempting to find mechanical explanations and interpretations of the vital process of plants, the nutrition, sap-circulation, growth, and movement.

It is true that the ultimate sources of vital phenomena remain unrevealed to us. And yet the desire to represent all processes as effects, and then to trace the causes of those effects, represents the driving force at the very root of modern research.

On viewing the garden, the mere fact of linking facts together causes the fascinating creation of ideas involving interdependence among the phenomena observed. The more imaginative an investigator, the more keenly he is impelled to discover relationships under the mute riddles that are presented to us in the form of plants. He becomes fascinated by the twin aspects of sexual reproduction as an aspect of natural selection and as an aspect of experimental propagation which is the result of experiments by gardeners.

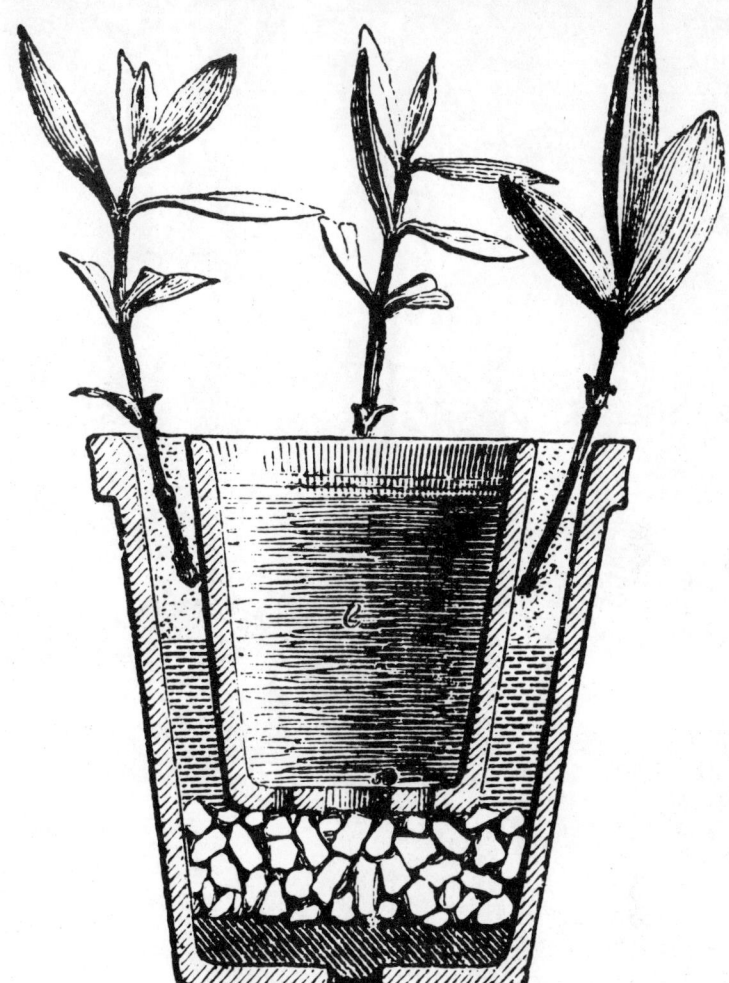

STRIKING CUTTINGS.

Modern science is governed by a desire to lay bare the causes of all phenomena. Thus investigators are no longer satisfied with knowledge about the existence of cells. Now we also want to know what are the functions of the various bodies which are formed within the protoplasm. Why is the cell-membrane thickened at a particular spot in a particular manner? What is the meaning of all the tubes and passages which exhibit such great diversity of size and shape? What part is played by the peculiar mouths of these channels, and why do they vary so greatly in shape and distribution in plants which are subject to different external conditions?

For modern researchers there is no fact without significance. Not too long ago botanists were content to determine in what manner the rudimentary organ of a plant is produced. Now we want to know why one rudiment grows and develops while another is obliterated.

The various lines of botanical research, with their particular problems and objects, have very little connection with each other. Several of these lines received their greatest impulses from Darwin. But it remains true that all the different departments of botany are more or less limited to description.

And yet, even though we are sure that we shall never be able to fathom the truth completely, we still go on seeking it. The processes we see involve movements. And these movements are life.

Every one of our theories has its history. In the first place a few puzzling facts are observed, and gradually other facts come to be associated with them. It begins to appear that there is a general uniformity underlying the accumulation. Botanists attempt with only faltering success to embrace the uniformity, and then comes along the master mind, the Linnaeus, the Darwin. The facts are collated by the master. He generalizes and announces a solution to the prevailing puzzle.

But then newly-observed facts are uncovered which do not adapt themselves to the theory of the existing generalization, and the new facts are considered for a time to be exceptions to the rule. The facts grow in number and the prevailing theory begins to crumble, and events await a new master and a new generalization.

The gathering of relatively minor facts and observations by modest laborers in the gardens is a collection of activities as vital in their totality as are the efforts of the occasional master. The study of dried plants made by a student in a provincial museum, the discoveries by an amateur about the flora of some remote valley, the facts gleaned by farmers and foresters in their work, the facts wrung from plants in the university laboratories—all these results can be turned into account. "Prove all things; hold fast that which is good."

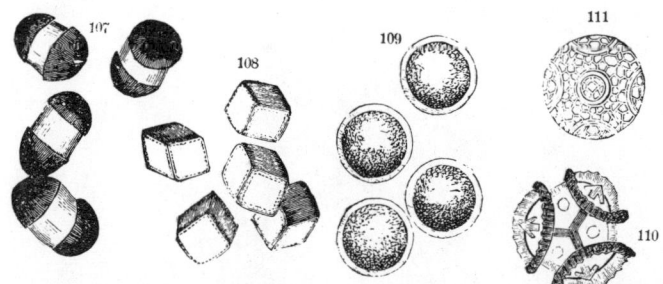

Pollen grains.—107, Pinus larico. 108, Basella rubra. 109, Ranunculus repens. 110, Scolymus grandiflorus. 111, Passiflora incarnata.

PROPAGATION

PROPAGATION BY SEED

Most plants used by man are propagated from seeds. For that reason alone one should be familiar with the development and behavior of seed activity.

Seed plants (spermatophytes) are divided into two main groups. The evergreens (conifers) which include pines, yews, spruce, hemlock and junipers have naked seeds and are called gymnosperms. The true flowering plants (oak, apple, elm, cherry, maple, flowering shrubs and vegetables) have their seeds enclosed in a fruit and are called angeiosperms. The seeds then develop from cones or true flowers.

TYPES OF SEED

From the standpoint of collection and extraction, seeds fall into three groups:

1. True seeds, readily extracted from dry fruits or cones. This group includes most conifers (fir, hemlock, larch, pine) and species bearing dehiscent fruits such as pods (honeylocust, locust, yellowwood) and capsules (fremontia, poplar, willow). Extraction of the seeds from such fruits usually involves drying by solar or artificial heat, followed by threshing or shaking. Commercial seed is almost invariably the true seed.

2. Dry fruits, with seed surrounded by a tightly adhering pericarp. These are of three main types: achenes, free of appendages (eriogonum), or retaining the feathery styles (clematis, cliffrose); nuts (chestnut, filbert, oak); and samaras, or key fruits (ash, elm, maple).

Other types of dry fruit used commonly by gardeners include:
A. Dehiscent Fruit
  1. Legume—bean, peanut (split on both seams)
  2. Follicle—peony and larkspur (split on one seam)
  3. Capsule—tulips, lily (split in various ways)
  4. Silque—cabbage (separate at maturity)
B. Indehiscent Fruit
  1. Achene—one seed—sunflower
  2. Caryopsis—one seed—rice and corn
  3. Samara—Maple
  4. Schizocarp—dill & celery
  5. Nut—Oak, hickory

Seeds of this group are rarely extracted from the fruit. For practical purposes the entire fruit is the seed, although sometimes the styles of some achenes and the wings of samaras are removed to reduce bulk and facilitate handling.

3. Fleshy fruits, such as accessory fruits (buffaloberry, wintergreen); aggregate fruits (raspberry); berries (barberry, currant, honeysuckle); drupes (cherry and plum, dogwood, walnut); multiple or collective fruits (mulberry, osage-orange); and pomes (apple, pear). In all cases, the seeds can be easily extracted by macerating the fruit in water and allowing the fleshy pericarp to float away, or, as in the walnut, by removing the pericarp or husk in a corn sheller. Sometimes fleshy fruits are dried and sown without extracing the seeds. Seeds also may be removed by maceration from the berry-like cones of juniper and the fleshy arils of yew.

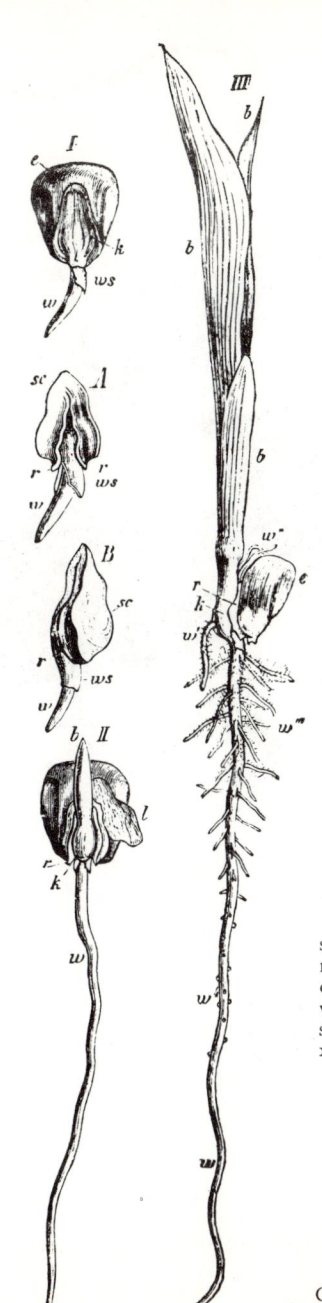

INDIAN CORN

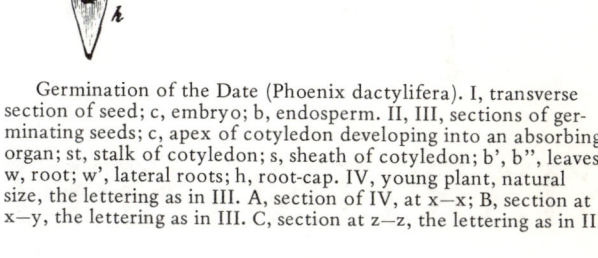

DATE

Germination of the Date (Phoenix dactylifera). I, transverse section of seed; c, embryo; b, endosperm. II, III, sections of germinating seeds; c, apex of cotyledon developing into an absorbing organ; st, stalk of cotyledon; s, sheath of cotyledon; b', b'', leaves; w, root; w', lateral roots; h, root-cap. IV, young plant, natural size, the lettering as in III. A, section of IV, at x—x; B, section at x—y, the lettering as in III. C, section at z—z, the lettering as in III.

Germination of Indian corn. I, II, III, successive stages. A and B, front and side views of a separated embryo. In the figures, w, the primary root; ws, its root-sheath; w', w'', adventitious roots; w''', lateral roots springing from the main root; e, part of seed filled with endosperm; sc, cotyledon; r, its open margins; k, the plumule; b, b', b'', leaves of young plant; l, fragment of wall of ovary.

STARTING SEEDS

Top left. Use sterile containers and planting medium—sterilized soil or vermiculite.

Top right. Press the moist planting medium firmly in the container.

Bottom. Tap the seed packet with your forefinger to distribute the seed at the rate recommended on the label.

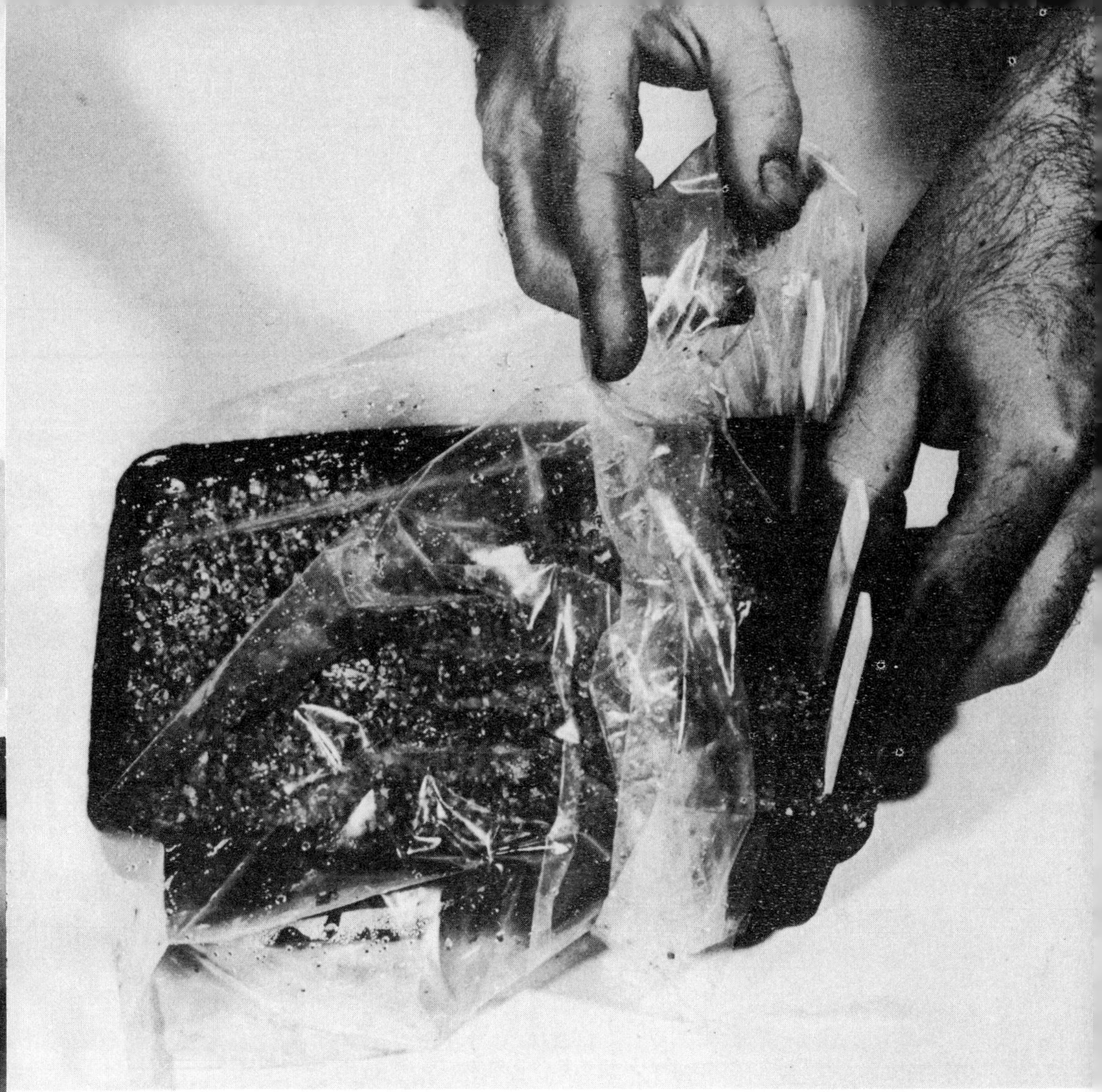

Top left. Wet the seeded container until water runs out of the bottom.

Left. Cover large seeds with a layer of fine vermiculite. Leave small seeds uncovered.

Top right. Place the seeded container in a polyethylene bag and keep it in a warm place until the seeds germinate. Then remove the bag and begin watering and fertilizing the seedlings.

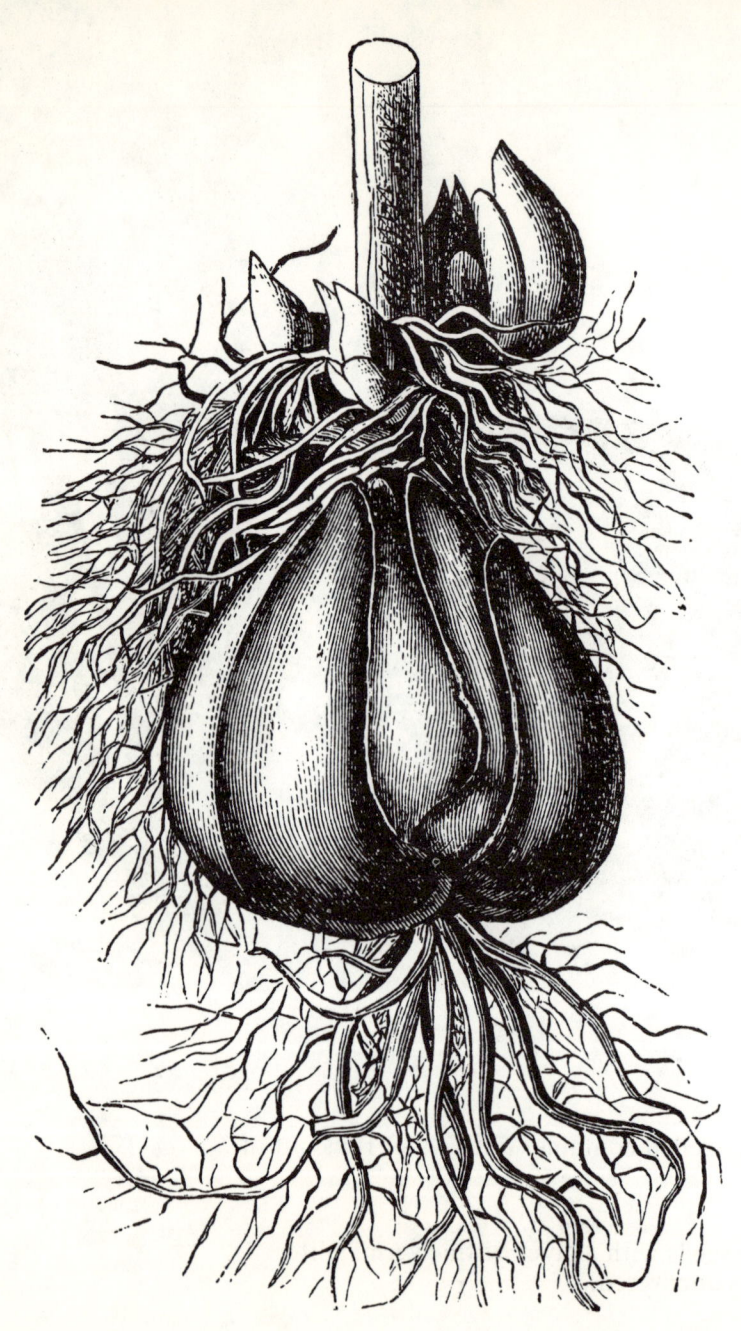

LILY BULB WITH BULBLETS

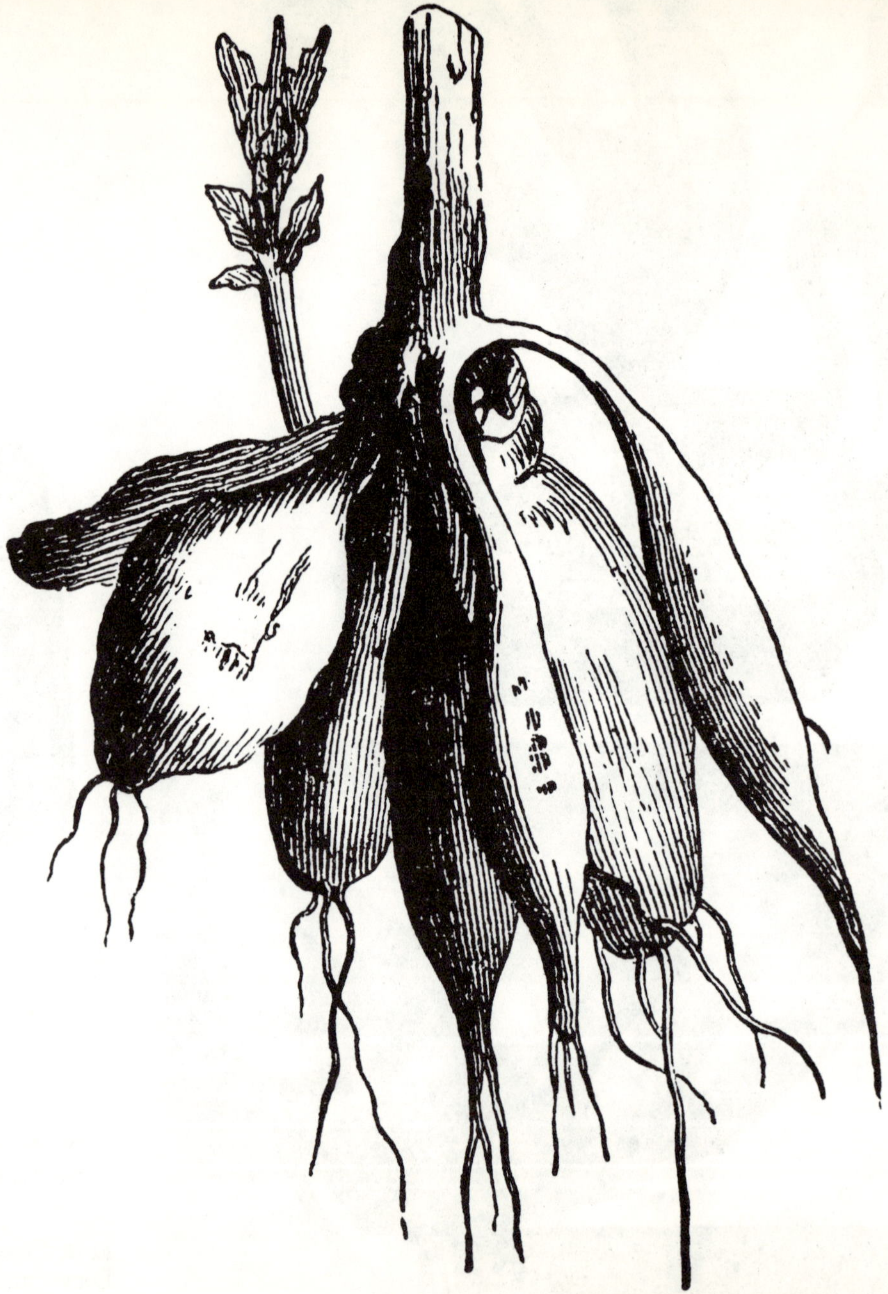

DAHLIA—TUBEROUS ROOT

PROPAGATION BY BULBS, CORMS, AND TUBERS

A Bulb is composed of either modified leaves in the form of scales, as in Lilies, or the bases of ordinary leaves folded round each other, as in Crinum, and held together by a more or less flatened axis, in the center of which is the growth bud or buds which never elongate, the flower stem being produced separately from the base of one of these buds. The best illustration of this manner of growth is seen in the bulb of Lilies. So long as the growth bud remains solitary, only one bulb is formed, but when more buds are developed the bulb divides into several. Some bulbs rarely multiply in this way, while others do so very freely. For those bulbs which do not divide, artificial means are resorted to for the purpose of multiplication. The central bud is cut out, or destroyed with a pointed stick, and this causes the bulb to develop lateral buds. Or the base is divided into four or more pieces, and this results in the formation of numerous bulbils. Choice Hyacinths are largely propagated in this way, in fact any true bulb that is strong enough to bear the treatment. Bulbs with scales, such as Lilies, may be propagated by breaking off the scales and pricking them separately into pans of sandy soil. Most Lilies may be readily and abundantly increased in this way. Some Lilies form bulbils in the axils of the leaves, and these may be removed and planted.

Many bulbs develop offsets from the base, as, for instance, in Narcissus, which may be removed for purposes of propagation.

A Corm is a short, solid, fleshy, more or less conical stem on which roots grow, etieher from the base only, or from all parts; the buds also may be scattered in like manner. Some buds grow into new corms which supplant the old one, as in Crocus, Gladiolus, etc. Nearly all corms multiply themselves freely, and it is not therefore often necessary to do more than remove the young offsets to grow them. Gladiolus produces numerous small basal corms called "spawn"; they also develop clusters of small corms on the flower stems.

A Tuber is a short thickened rhizome or stem, bearing buds and node-like scars, the best examples being the Jerusalem Artichoke, the Potato, and the Yam. The flesh subterranean growths of the Dahlia are not true tubers, but simply fleshy roots, as they do not bear buds, but are reservoirs. The rootstock of some Nymphaeas and of Nelumbium are tubers. Propagation by means of tubers is simply stem division, and wherever a bud can be severed with a portion of the fleshy stock, it may be utlized.

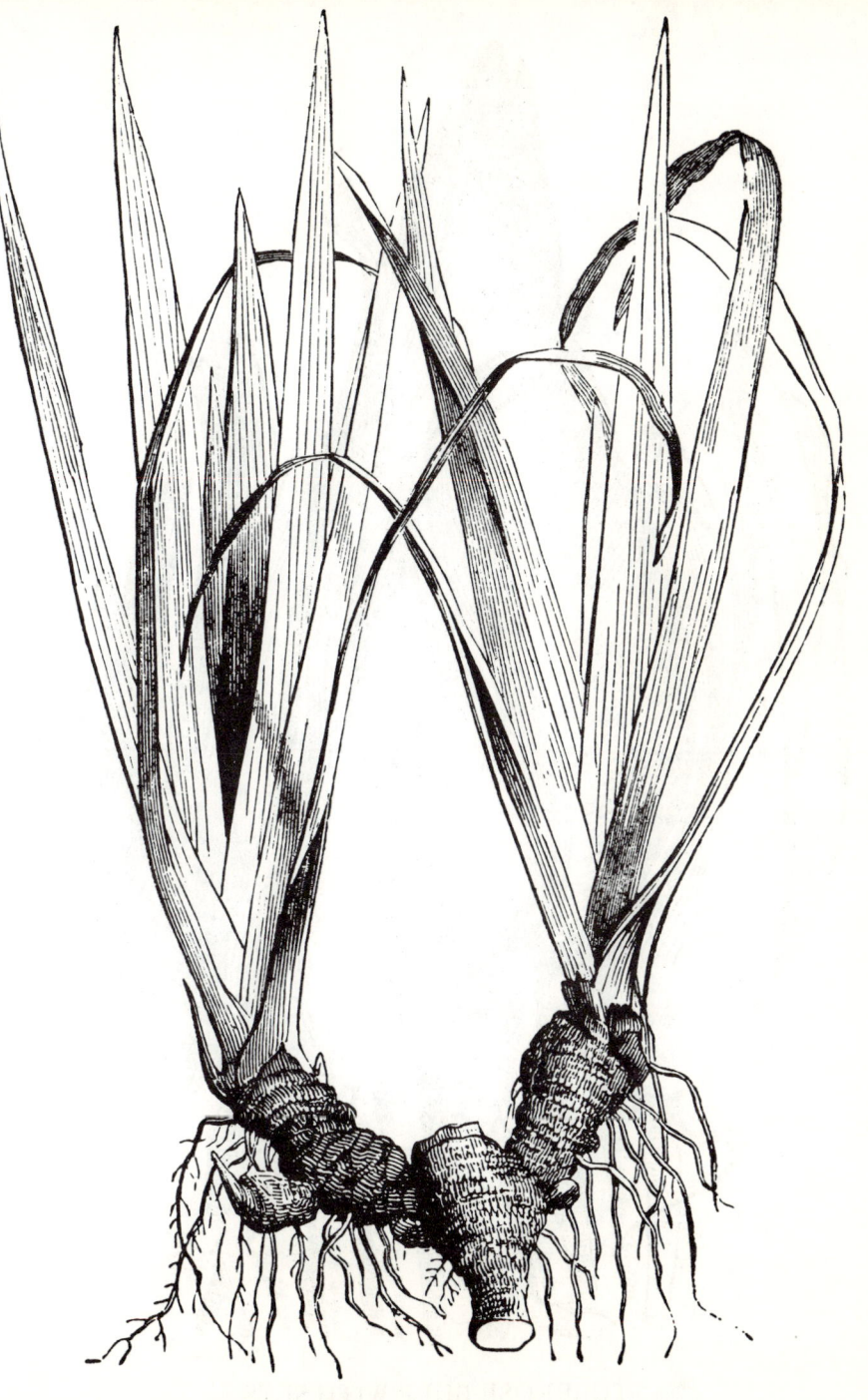

Rhizome of Iris

BULB OR CORM OF GLADIOLUS (UPPER PART).

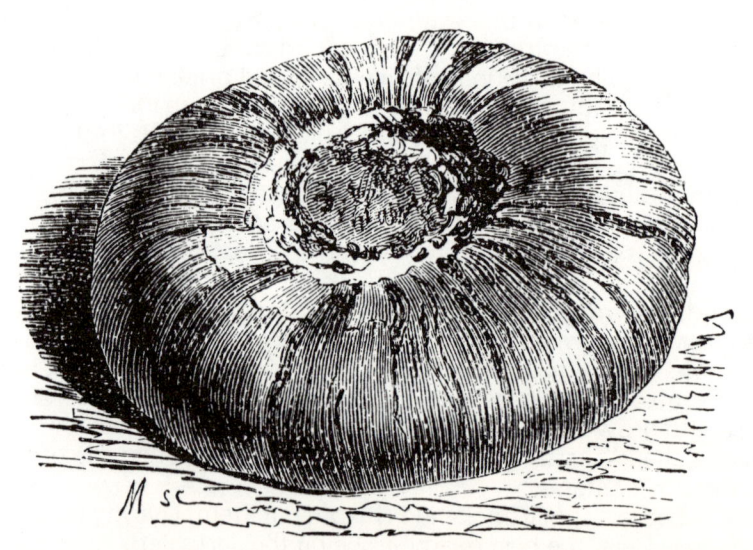

BULB OR CORM OF GLADIOLUS (UNDER PART)

TUBEROSE BULB WITH SETS

FORCING BULBS

Bulbs can be forced to bloom indoors earlier than they normally would outdoors in the garden or yard. The easiest bulbs to force are crocus, galanthus, hyacinth, narcissus, scilla, and tulip. A nurseryman can tell you the varieties that are best suited for forcing.

Forcing bulbs includes two phases. The bulbs develop buds and roots in the first phase and bloom in the second.

You should begin the first phase in October or early November. Plant the bulbs in pots and keep them at a temperature of 40 degrees F. for 8 to 12 weeks. During this phase, you can keep the potted bulbs outdoors or in a cold room indoors.

If you keep your bulbs indoors, the room must be dark and kept at 40 degrees F. Do not let the soil in the pots dry out; water the bulbs every day.

The second phase begins about mid-January after shoots have appeared on the bulbs. When the shoots are well out of the necks of the bulbs, bring the bulbs into a cool, bright room that can be kept at 55 degrees F. They will bloom in about 1 month.

You may refrigerate crocus, hyacinth, narcissus, and tulip bulbs at 40 degrees F. for 2 months instead of planting them in pots. At the end of 2 months, plant the bulbs in bowls and start them in the second phase of development.

You should discard bulbs that you force. They seldom grow and flower well when replanted in the garden.

THE TULIP

STEPS IN FORCING BULBS

Place the potted bulbs in a cool, bright room. Keep them at a temperature of 55 degrees F. until they bloom. Water the bulbs daily.

A. Clean the pot and cover the drainage hole with a clay plug.

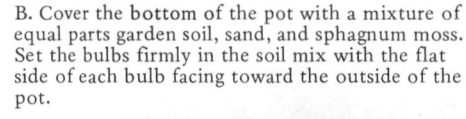

B. Cover the **bottom** of the pot with a mixture of equal parts garden soil, sand, and sphagnum moss. Set the bulbs firmly in the soil mix with the flat side of each bulb facing toward the outside of the pot.

C. Cover the bulbs with the soil mix. Press it firmly around and over the bulbs.

D. Place the pots in a stone well around a tree and cover them with a layer of loose leaves or straw. Never pack the leaves or straw because water must drain freely through them.

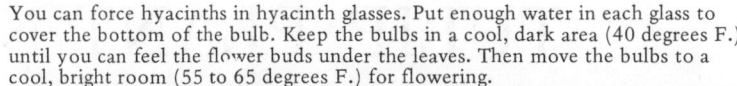

You can force hyacinths in hyacinth glasses. Put enough water in each glass to cover the bottom of the bulb. Keep the bulbs in a cool, dark area (40 degrees F.) until you can feel the flower buds under the leaves. Then move the bulbs to a cool, bright room (55 to 65 degrees F.) for flowering.

E. If you do not have a stone tree-well, you can bury the potted bulbs in a pit. Set the pots close together and cover them completely with soil. Put a wire screen over the pots to protect the bulbs from rodents, moles, and other animals.

F. You also can force bulbs in vermiculite. Follow the same steps that you would if you were using a soil mix.

G. Inspect your bulbs occasionally. When the shoots are well out of the soil, bring the bulbs into a cool room for flowering. The shoots will be pale green to almost colorless.

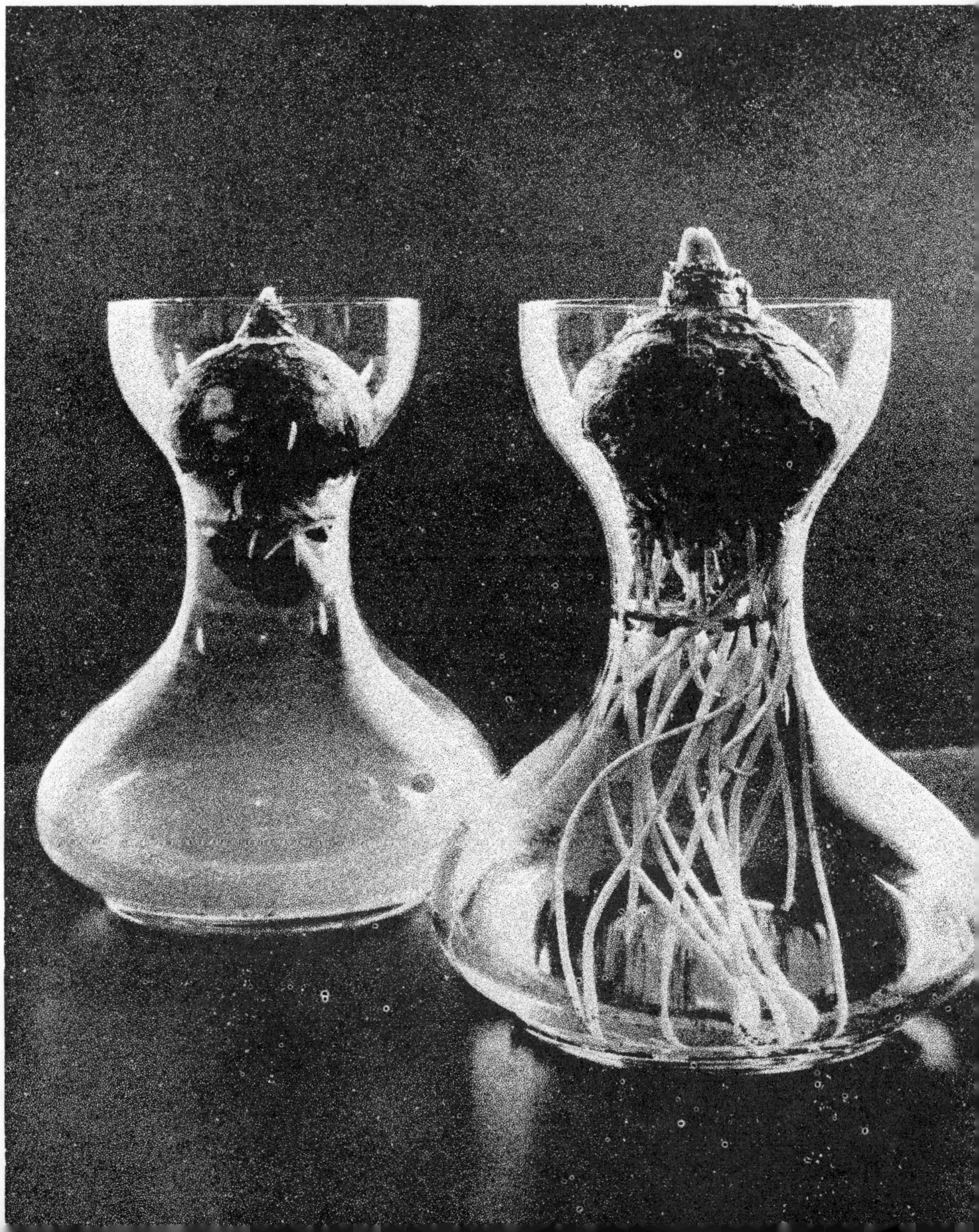

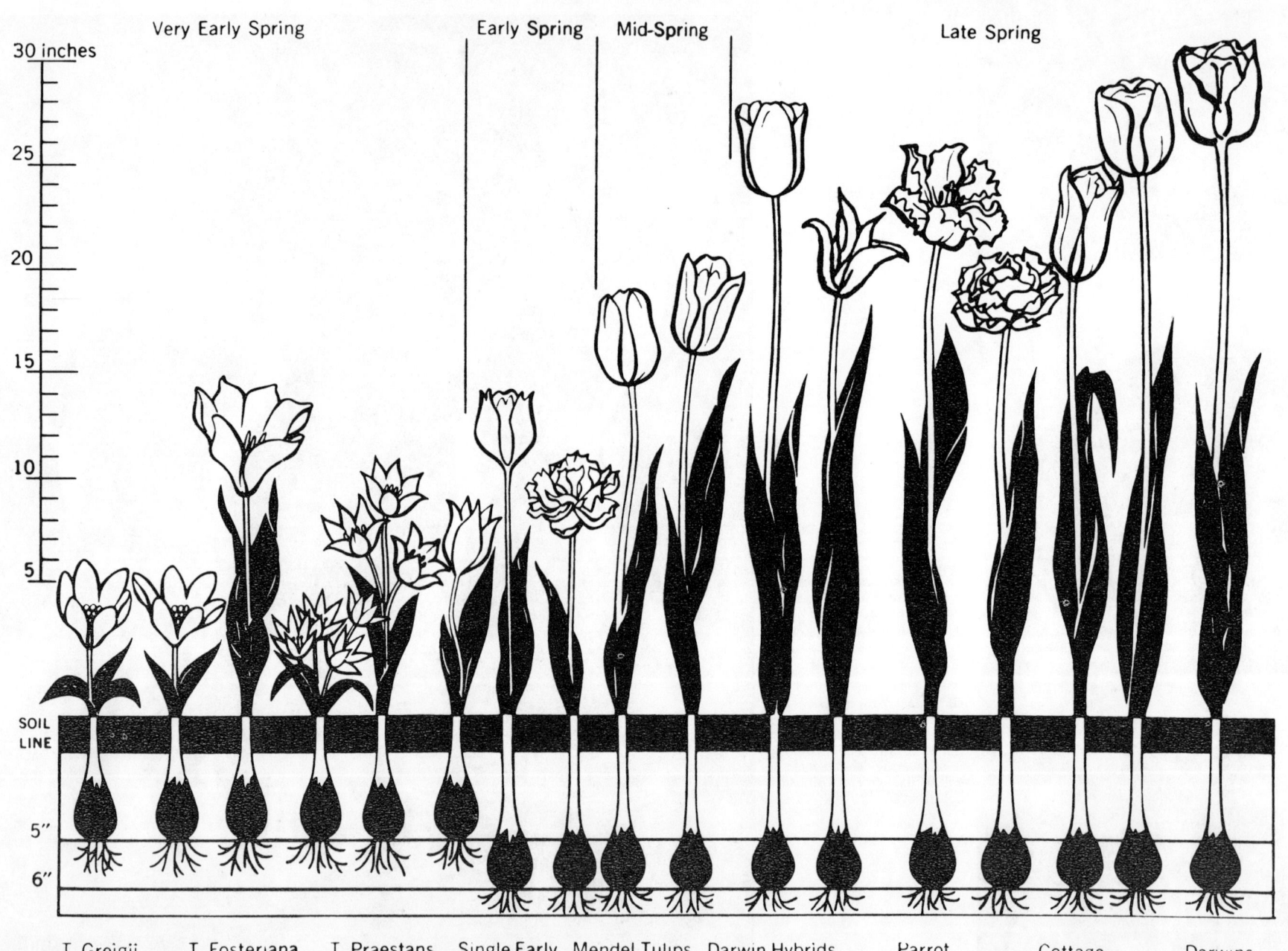

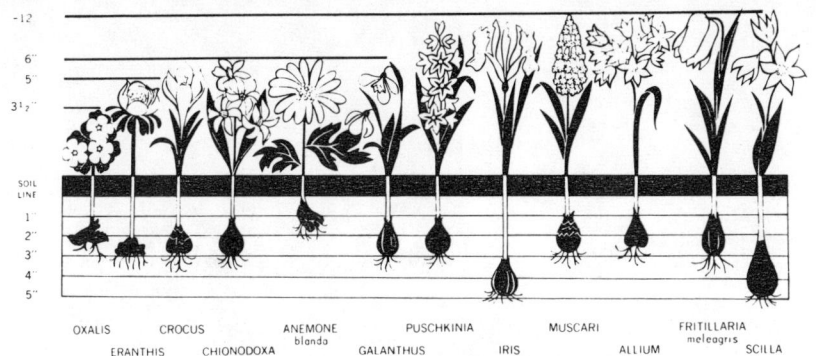

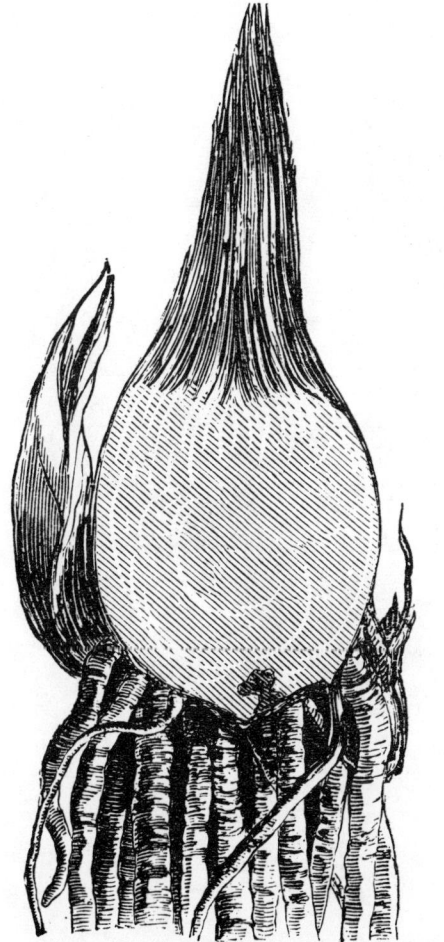

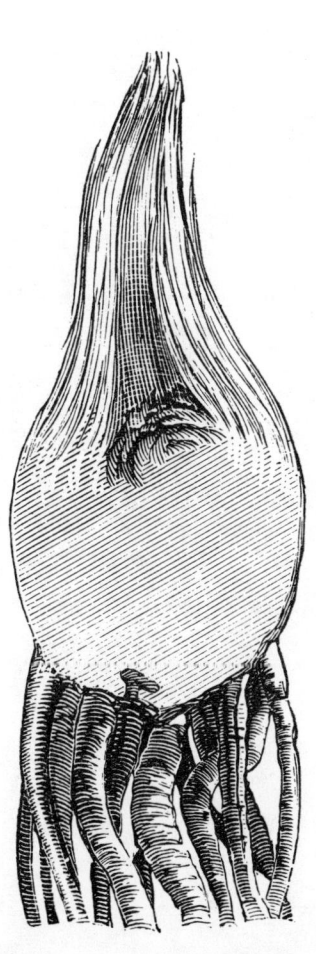

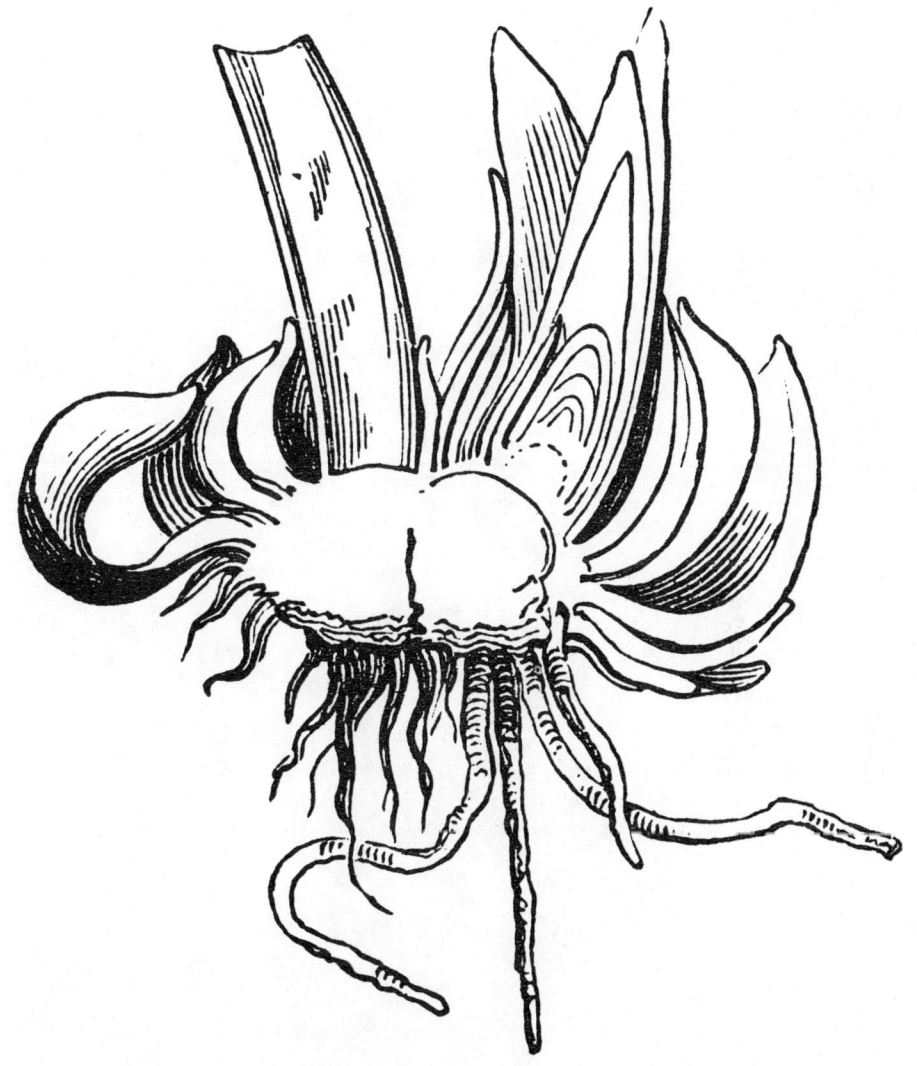

SOUND BULB **BULB DECAYED AT CENTER**

Lily—Scaly Bulb

LAYERS

If a branch of an ornamental plant is wounded and the wound is covered with a rooting medium—soil or sphagnum moss—the branch usually will strike roots while it is still attached to the parent plant. It then can be severed from the parent and set out as a new plant. This method of propagation is layering. It is successful with more species of trees and shrubs than is any other method of vegetative propagation.

Layering usually is most successful if done in spring or in late summer; rooting is most vigorous in cool weather.

If a branch is low and sweeping and can be bent to the ground easily, layer it by burying the wounded part in soil. This is called simple layering. If the branch cannot be bent to the ground, layer it by wrapping the wound with moistened sphagnum moss. This is air layering.

Simple Layering

Professional propagators use many variations of the layering-in-soil method of propagation. The easiest for the home propagator is simple layering—burying a single branch in the soil with only its tip protruding.

Before making a simple layer, work leafmold or peat and sand into the soil where the branch will be layered.

Begin the layering operation by wounding the branch. Make a slanting cut 2 inches long on the upper side of the branch about 12 inches from the tip. Dust the cut with rooting stimulant.

Then fasten the branch to the soil. Pin it down between the trunk and the cut with a wooden peg or wire wicket, or weight it with a stone.

After the branch is pinned to the soil, bend the tip upright. As you do this, twist the branch as if you were turning a screwdriver one-half a turn. This will open the cut.

Next, place a second peg or pin over the branch directly at the point of the cut.

Cover the pegged branch with several inches of soil into which leafmold or peat and sand have been worked. Mound the soil around the upturned stem so the wound is 3 or 4 inches underground. Pack the covering soil firmly.

Mulch the soil over the layered branch with straw or leaves. Water frequently; keep the covering soil moist.

When the layer has formed roots—the following spring for spring-layered branches, or the second spring for fall-layered branches—cut the rooted branch free from the parent plant.

Leave the new plant in place for 2 or 3 weeks after it is severed from the parent.

This will give it time to recover from the shock of being cut. Then transplant it to a nursery bed, where it should be tended carefully for a year.

HOME PROPAGATION OF TREES AND SHRUBS

Many kinds of ornamental trees, flowering shrubs, roses, and evergreens are easy to propagate by home methods. These home methods utilize inexpensive, easily available materials and equipment.

Plants propagated by layers, cuttings, or grafts have characteristics exactly the same as the parent plants. Plants propagated by seeds often have characteristics different from those of their parents.

Many ornamental trees and shrubs can be propagated by layering. Because a branch of the parent plant is needed to form each new plant, this method is practical only for propagating a small number of plants.

Many kinds of trees and shrubs can be propagated from cuttings. One parent plant yields enough propagating material to start a large number of new plants. The new plants are small, however, and must be tended carefully for several years.

Some trees and shrubs can be propagated by grafting. You can propagate a large number of new plants from a small amount of propagating material by bud grafting. Or you can get large new plants soon after propagating by cleft grafting. Bud grafting and cleft grafting demand more skill and specialized knowledge than the other methods of propagation.

For greatest success in propagating—
USE ONLY HEALTHY PLANTS.
PROPAGATE IN THE PROPER SEASON.
PROTECT PROPAGATING MATERIAL FROM DRYING.
GIVE NEWLY PROPAGATED PLANTS EXTRA CARE UNTIL THEY ARE WELL ESTABLISHED IN THEIR PERMANENT LOCATION.

Bush.

Rooting Stimulants

Several organic chemicals stimulate the formation of roots on layers and cuttings. Preparations containing these chemicals are available from garden-supply stores.

Some brands of rooting stimulant are sold in several strengths.
Follow directions on the container label when using rooting stimulants.

Air Layering

Kits containing all the materials needed for air layering are available at garden-supply stores. If you do not buy a kit, you will need the following:
A SHARP KNIFE.
ROOTING STIMULANT.
SPHAGNUM MOSS.
POLYTHENE PLASTIC SHEETING.
PLASTIC ELECTRICAL TAPE.

One-year-old branches are best for air layering. Older branches may form roots, but take longer to do so than branches that are 1 year old.

Make the layer 12 to 18 inches from the tip of the branch. If there are any leaves within 6 inches of the point where the layer is to be made, remove them.

Begin air layering by wounding the branch. Make a shallow, slanting cut about 2 inches long in the branch. Dust rooting stimulant into the cut and place a small sliver of wood in the cut to keep it open.

Next, dampen a fist-size ball of sphagnum moss and squeeze it as tightly as possible to remove excess water. Wrap the sphagnum around the branch, covering the wound.

Then, cover the ball of moss with plastic sheeting—an 8- by 10-inch sheet is big enough. Wrap the plastic around the moss-covered branch so the sheet overlaps itself.

Finally, twist the ends of the plastic sheeting around the branch and fasten them securely with plastic electrical tape.

Watch for collection of rain water in the plastic package. If water seeps in the ends of the package, punch a small hole in the plastic to allow the water to drain out.

Leave the air layer undisturbed for one full growing season. Branches that are layered in the spring should root by the following spring. Branches layered in the fall should root by the second spring. When roots have formed, they usually are visible through the plastic covering.

In spring, after the roots have formed, remove the plastic and cut off the branch below the roots. Set the new plant in the nursery bed.

If roots have not formed by the third spring, remove the wrappings from the branch. Neither the parent plant nor the branch will be seriously harmed if the layer is unsuccessful. Try layering another branch; you may be successful this time, even though your first attempt failed.

Care After Rooting

The root systems of newly rooted layers are small in relation to the tops. You can reduce loss of water through the leaves of these new plants—and lessen danger of wilting and plant death—by pruning and shading.

As soon as you plant new rooted layers in the nursery bed, prune all side branches; remove one-third of their original length.

Erect screens around the new plants to shade them. Suitable screens can be made of snow fencing, lath, reed matting, or burlap attached to wood framing.

After the first winter remove the screens. The roots should be large enough by the end of the season to absorb all water needed by the plants. The plants then can be safely transplanted to their permanent location.

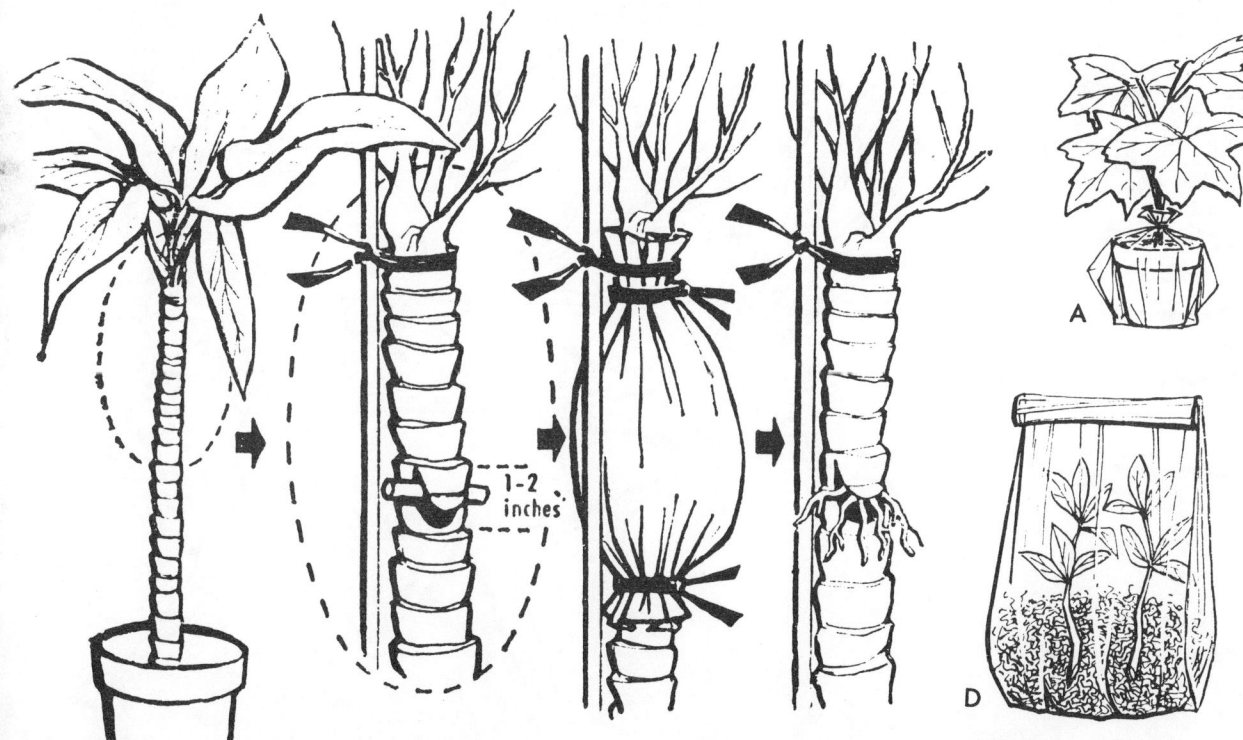

Use of polyethylene films. A. Cover clay pot of well-watered plant with polyethylene film. B. Cover propagation pan to keep humidity high around rooting cutting. C. Cover recently rooted, air-layered plant to retard water loss from foliage. D. Wrap unrooted cuttings in sphagnum moss and cover with perforated polyethylene film.

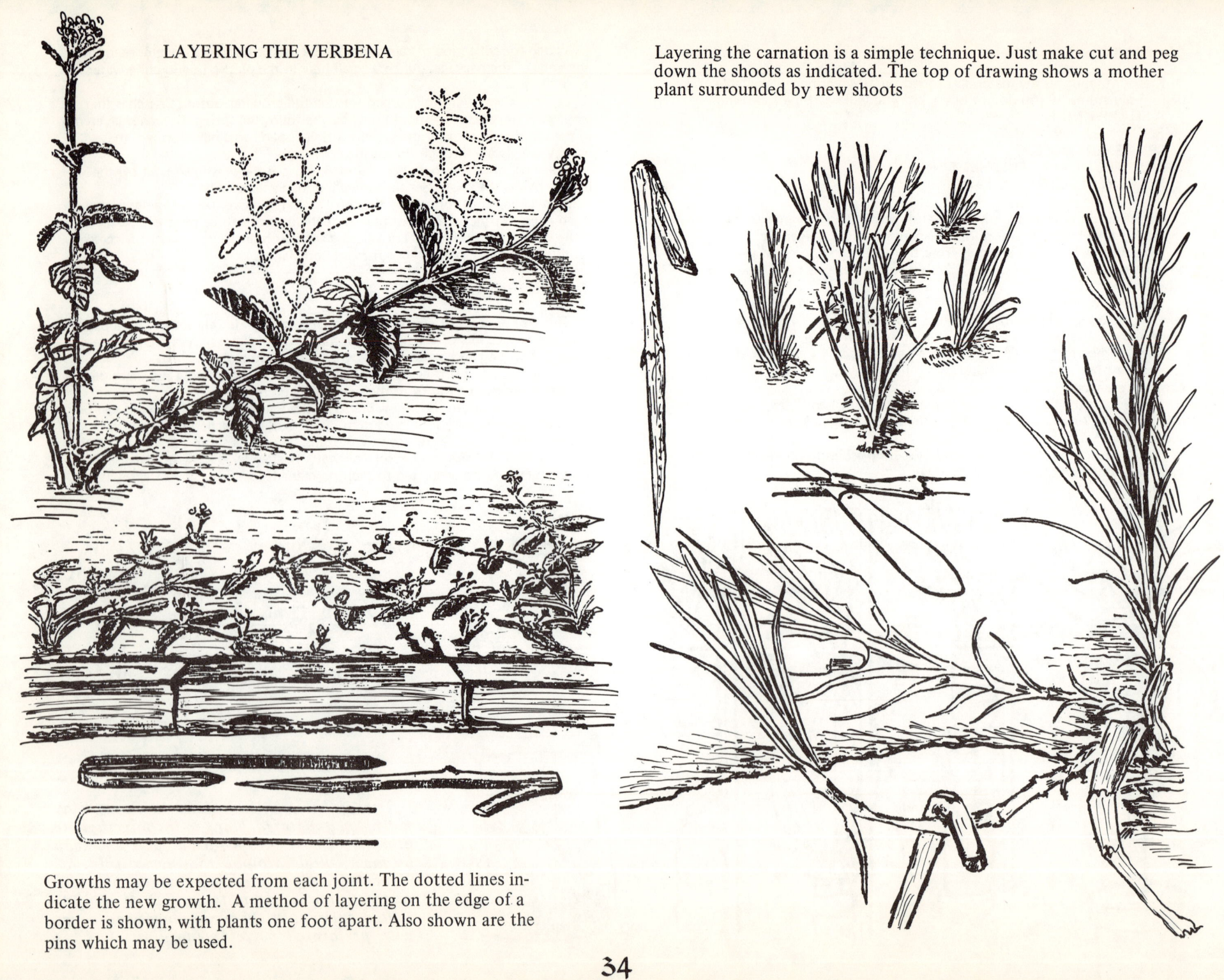

LAYERING THE VERBENA

Layering the carnation is a simple technique. Just make cut and peg down the shoots as indicated. The top of drawing shows a mother plant surrounded by new shoots

Growths may be expected from each joint. The dotted lines indicate the new growth. A method of layering on the edge of a border is shown, with plants one foot apart. Also shown are the pins which may be used.

How to Dwarf a Dracaena

Cut a notch around the stem of the plant directly under the main foliage spread. Cut a plastic pot in half and fill each half section with soil. Fit the sections of the pot around the notch cut in the stem of the dracaena. Bind the pot together with tape or cord. After the root sections begin to grow out of the notched area, cut from the larger plant. The stem will continue to produce cutting from the top down.

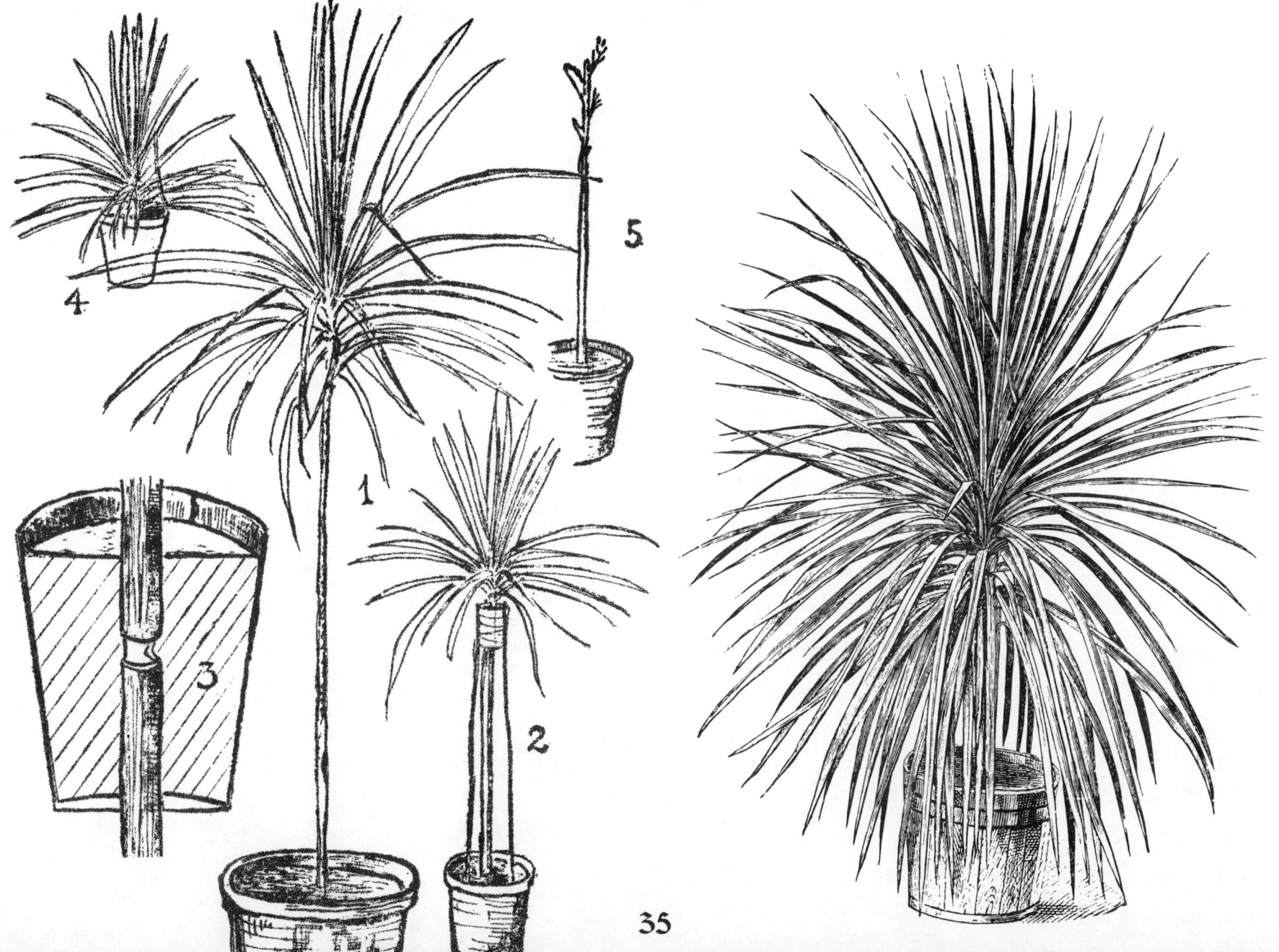

CUTTINGS

Cuttings, or slips, made of newly formed wood are easiest to root.

Take cuttings from roses and spring-flowering shrubs in midsummer when the new stems are no longer succulent but have not yet become hard.

Cuttings of some evergreens—holly, yew, arborvitae, and juniper—root best if they are taken from the plants in late fall or early winter, after they have been subjected to several heavy frosts. Boxwood cuttings can be taken at any season.

One way to root the cuttings is in a flower pot that is kept covered with a plastic bag. The plastic cover allows the cuttings to "breathe" but prevents loss of water.

For a rooting medium, use a mixture of 1 part clean sand and 1 part peat moss. Moisten the mixture. When the mixture has the proper amount of moisture, only a drop or two of water will come from a handful that is squeezed tightly. If you get the mixture too wet, add dry sand and peat to it.

Fill the flower pot with this rooting medium.

Now make the cuttings. Make a slanting cut through the stem 2 to 6 inches from the tip of the cutting.

Strip the leaves off the lower half of each cutting and dip the base of the cutting in rooting stimulant. Insert the cutting to about half its length in the rooting medium.

Put the cuttings close together; a 6-inch flower pot will hold 10 or 12 cuttings. When the flower pot is full, spray the cuttings lightly with water.

Now place the flower pot inside a polythene freezer bag. Twist the top of the bag closed and fasten it with a rubber band. This forms a miniature plastic greenhouse that is vapor proof; the cuttings will need no more water until they are well rooted.

Set the cuttings in a window where they are exposed to daylight but never to direct sunlight. Heat from direct sunlight may kill the cuttings.

Rooting Period

Cuttings of most plants will form roots within 2 months. Cuttings made in midsummer should be rooted by fall; those made in winter should be rooted by spring. After they have been in the flower pot for 2 months, very carefully dig one of the cuttings and inspect it for rooting.

If no roots are fisible, replant the cutting, close the bag, and set the flower pot back in the window. Hold summer cuttings until spring, winter cuttings until early summer. Then inspect them again for rooting. Continue periodic inspection during the growing season—about once a month—until the cuttings root or until they turn brown or black, indicating death of the cuttings.

Herbaceous Cuttings

The above drawing shows the proper method of making herbaceous cuttings.

1. Common Geranium
2. Begonia
3. Coleus
4. Fancy Geranium
5. South African Daisy
6. Heliotrope
7. Chrysanthemum (C. frutescens)
8. Slipperwort (Calceolaria)
9. Bellflower (Campanula)

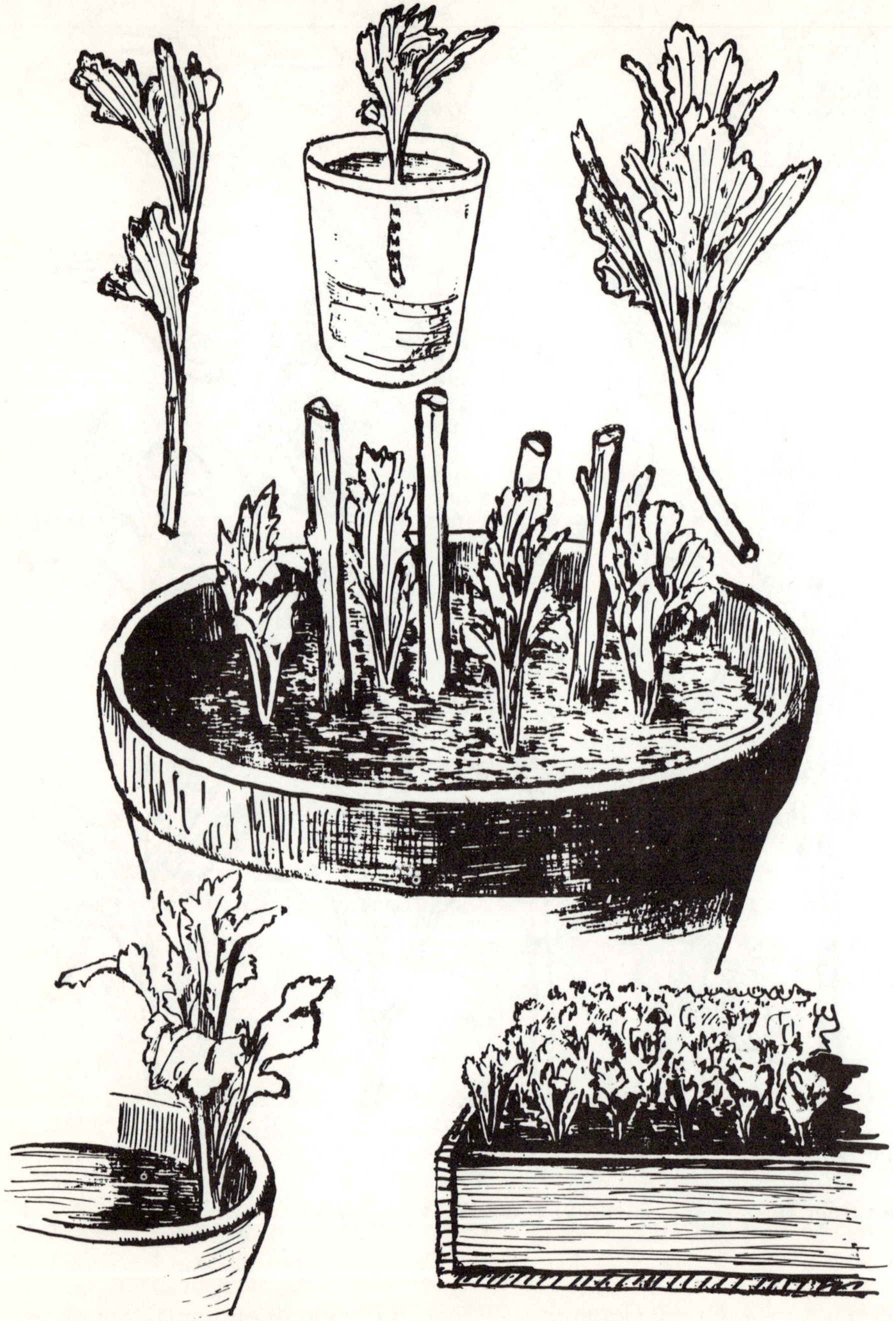

Chrystanthemum cuttings are easy and the only way to propagate the plant in a hurry. On the left top the cutting is too thin. On the right the foliage makes it a perfect cutting. The center pot shows a larger plant clipped and developing new shoots. You can either plant in single pots or flats.

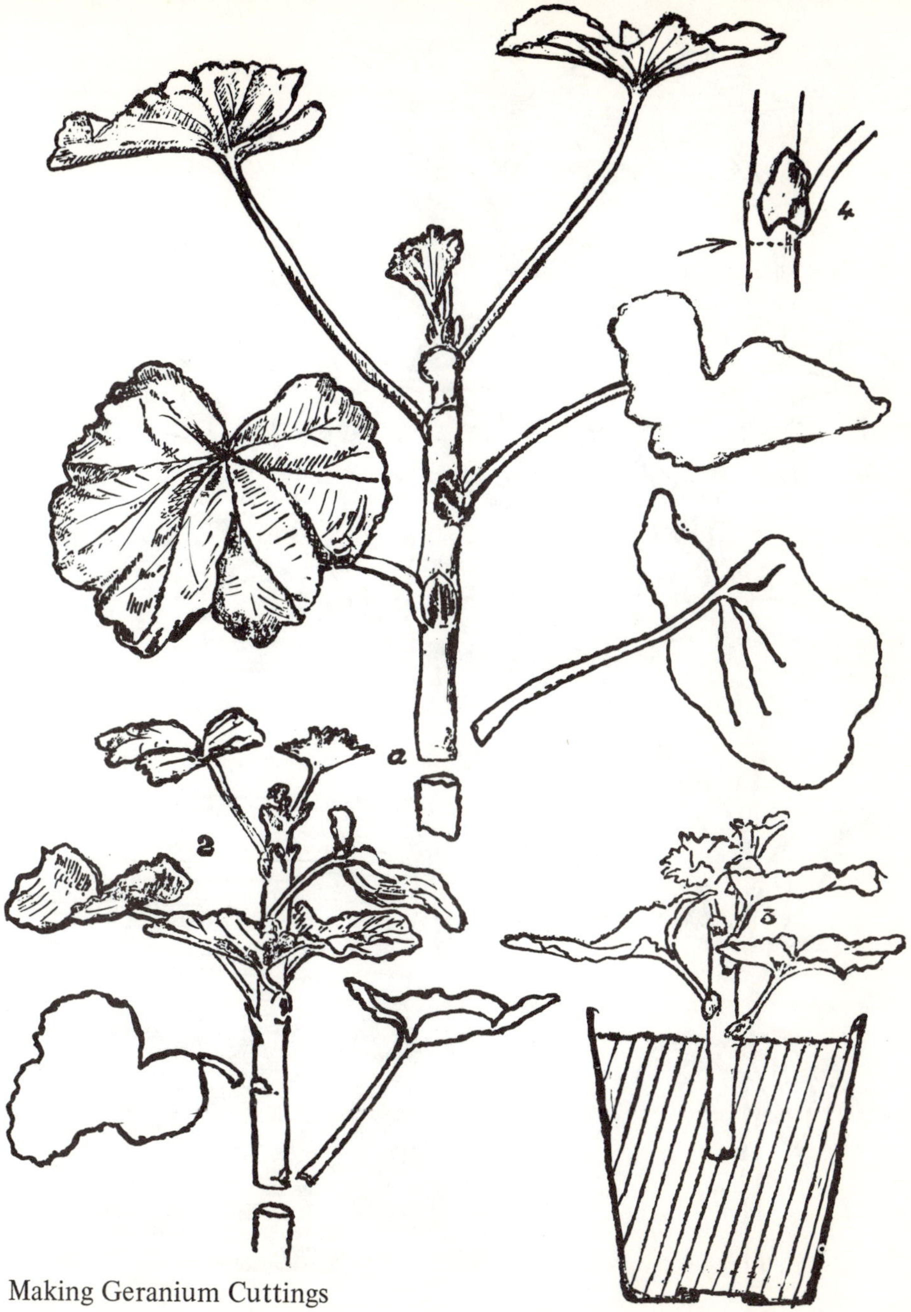

Making Geranium Cuttings

Follow a few simple directions. Remove one or two leaves from a heavily foliaged cutting. The cut should be made just below the stipule (indicating in drawing by arrow). Insert cutting in loose medium (a sandy mixed soil).

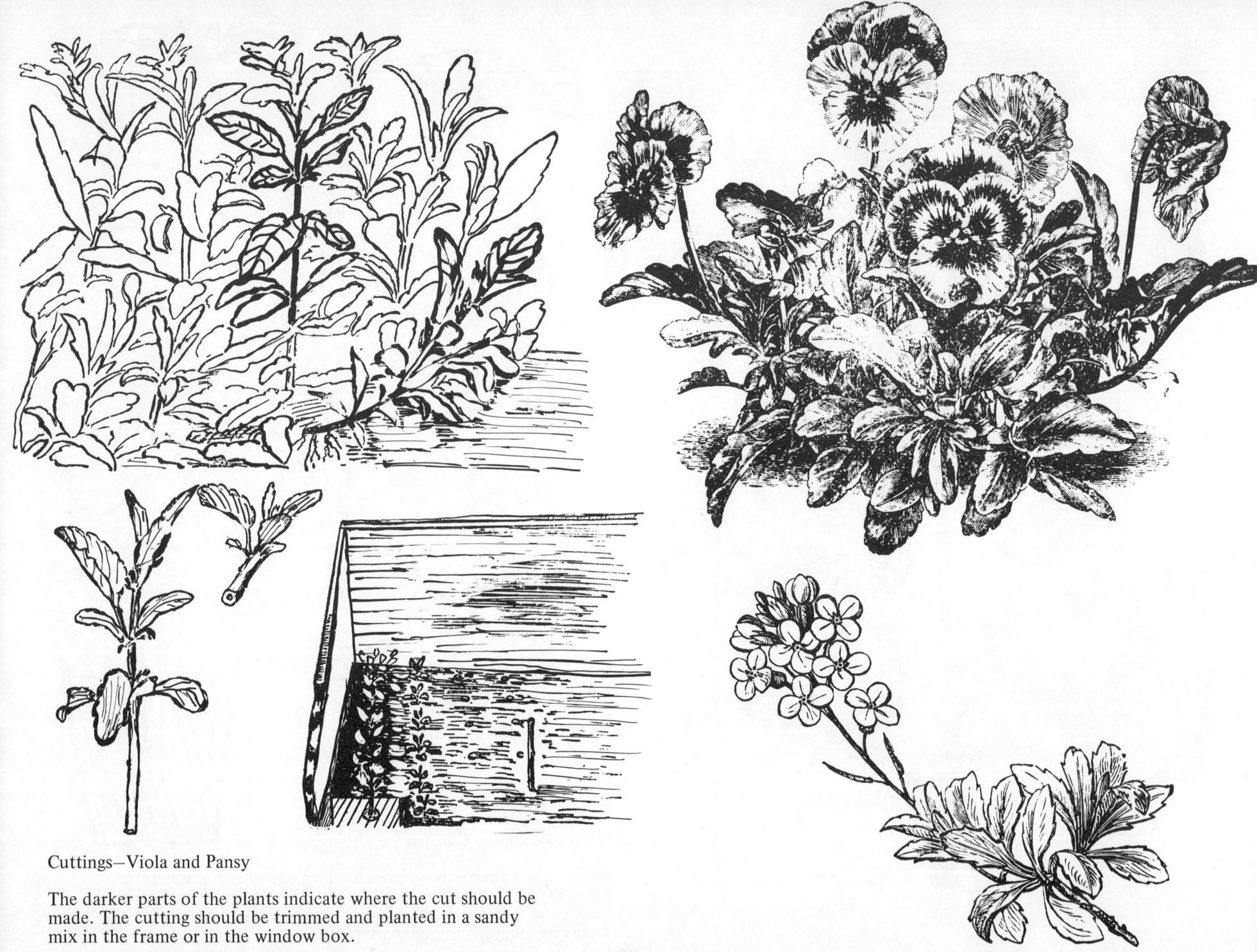

Cuttings—Viola and Pansy

The darker parts of the plants indicate where the cut should be made. The cutting should be trimmed and planted in a sandy mix in the frame or in the window box.

Care After Rooting

After cuttings have rooted, grow them in a coldframe for one winter before planting them in their permanent location. Harden the plants for moving to the coldframe by opening the plastic bag for an hour or two each day. After a week of this, the plants should be hardened enough to move safely.

If cuttings root in spring or early summer, transplant them immediately from the pot to an open coldframe. In fall, cover the coldframe with sash or plastic sheeting.

If cuttings root in late summer or fall, either transplant them immediately from the pot to a closed coldframe, or place the entire pot of cuttings in the coldframe and transplant in the spring.

If you decide to leave the cuttings in the pot, dig a hole in the soil of the coldframe and set the flower pot in the hole with its rim even with the soil surface. Fill in around the flower pot and firm the soil.

In spring, after they have over-wintered in the coldframe, move the plants to a nursery bed. Shade them and water frequently during their first season. The plants, except shade-loving kinds, do not need shading or special watering after the first year.

Transplant new trees and shrubs to their permanent locations after they grow 12 to 24 inches tall.

Freeze Damage in Coldframe

Heat from sunshine can warm the air inside a sash- or plastic-covered coldframe sufficiently to start growth in the new plants. If freezing weather follows, the plants may be killed by the cold.

To prevent this freeze damage, keep the plants cool and dormant. Place the coldframe in a shady location, or open the coldfame slightly on sunny days to ventilate it and dissipate the heat.

Span-roofed Frame

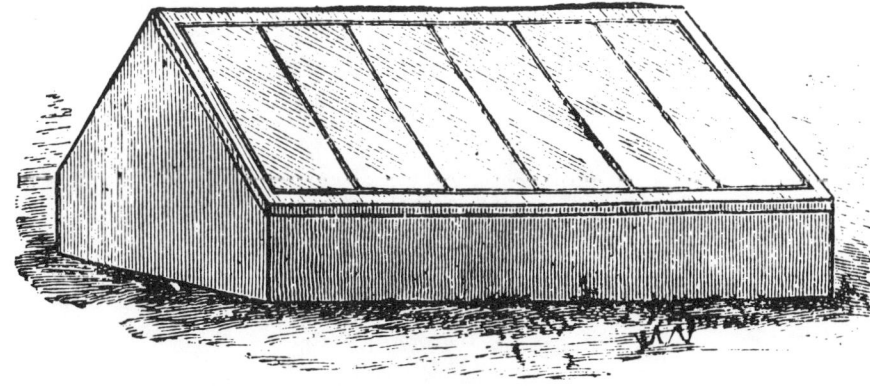

Evergreen Cuttings

1. Tea Olive
2. Yew, Taxus in variety
3. Skimmia
4. Speedwell, Veronica t.
5. Euonymus
6. Laurustiams, Virburnum tinus

PROPAGATION BY DIVISION

Strictly speaking all propagation, except by seed, is effected by division. If we propagate by cuttings, suckers, grafts, or buds, we must, in either case, divide the plant to obtain them. Propagation by division is, however, usually understood to imply the parting of tufted plants, so that each part has roots, and, if possible, growing points. The most favorable season for this operation is the spring. Many Orchids can only be multiplied by division. The same may be said of Bamboos, for which it is the only practicable means of increase, seeds being rarely available. Adiantum farleyense is a familiar example among many ferns, for the propagation of which division is the only means. It may be accepted as a general rule, that all plants of tufted habit, including even some Palms, may be increased by division. Herbaceous perennials that have annual stems can be divided with a spade or trowel. In some cases, a large number of plants can be obtained by division by introducing a quantity of fine soil among them, in order that the lower branches may strike root in it; or the plants may be taken up and replanted deeper.

PROPAGATION BY RUNNERS

Many plants, like the Strawberry, produce runners, which proceed along the surface of the ground, deriving nourishment from the parent plant, and develop later a bud on the upper side, and rudimentary roots. These, under favorable circumstances, root into the soil and assist in the extension of the parent plant. The point of the runner proceeds, and another plant is formed at the next joint or bud, and so on. In propagating by runners, if the object be to obtain as many plants as possible, the parent plants should be prevented from flowering. If particularly strong plants are required, the runner should be stopped after it has made one or two buds.

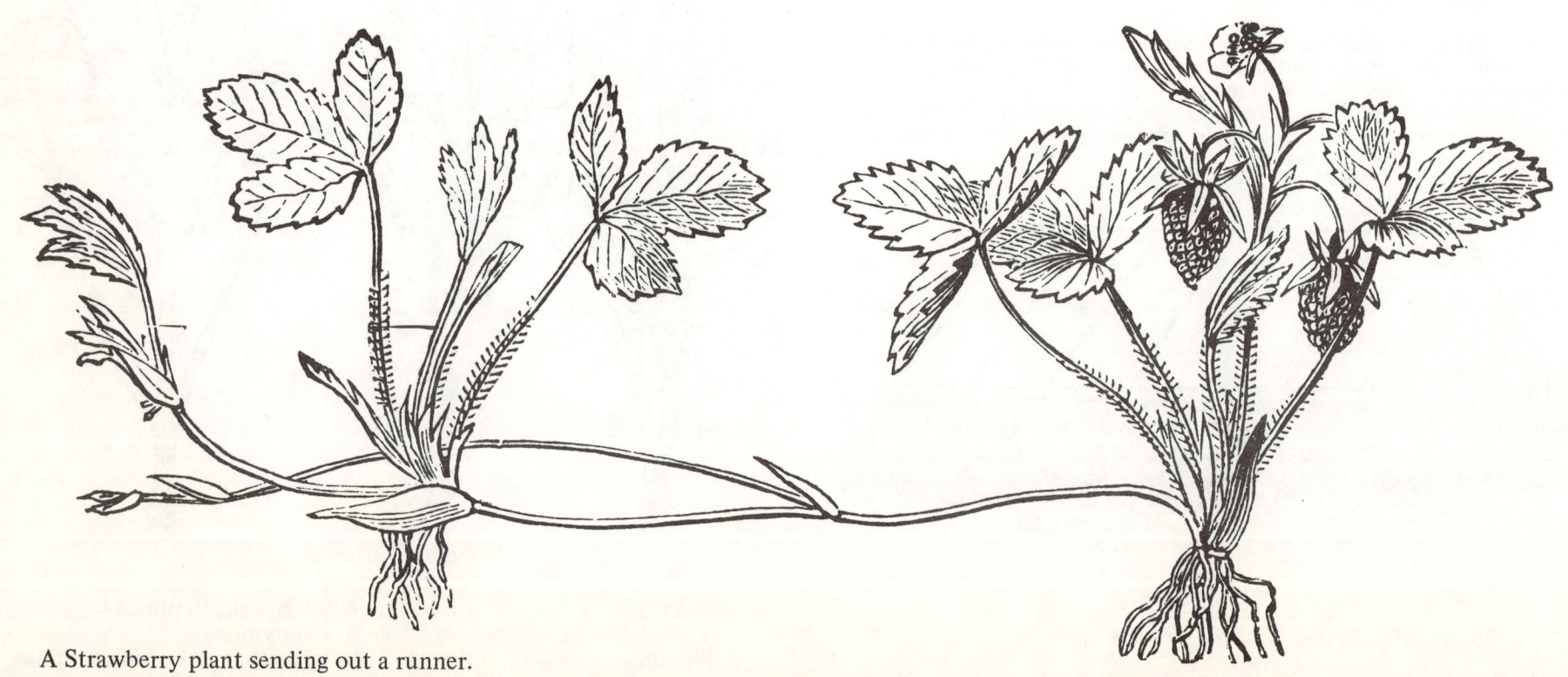

A Strawberry plant sending out a runner.

PROPAGATION BY ROOTS

All plants that readily throw up root suckers may be propagated by cuttings of the roots. Although the normal formation of growth buds is in the axils of the leaves, yet buds frequently appear irregularly on any part of the stem and roots. If healthy, vigorous roots of, for instance, the common Hawthorn, are chopped into short pieces, scattered on the surface of a piece of raked, dug ground, and then covered with soil, they will develop plants. Although cuttings of the roots will strike when laid horizontally, yet it is often better to plant them in an upright position, with their tops level with the surface. The cuttings may be from 3 to 9 inches in length; and in planting, care should be taken tha t the end which was nearest the stem be placed uppermost. The Plum, Apple, Pear, Quince, Rose, Robinia, Poplar, Elm, Mulberry, Maclura, Rhus, Calycanthus, Paulownia, and Sophora are some amongst the many trees which may be propagated from roots. Many herbaceous plants, as the Horseradish, Sea-kale, Anemone japonica, etc., may also be increased in the same way. It may be mentioned that a plant raised from a root cutting bears leaves, flowers, and fruit exactly similar to those of the original tree. For instance, trees have been rared from the roots of the Cortland, and they possessed all the qualities of that well-known Apple. Many Ferns produce adventitious buds on their roots. Davallia (rabbit foot) may be propagated by means of roots cut off and sown in pans of peat and chopped sphagnum.

PROPAGATION BY LEAVES

This mode of propagation is now often advantageously practiced for such plants as Gloxinia and other Gesneriads, Bertolonias, Begonias, Echeverias, Pinguicula, etc. There is no physiological reason why the leaves of all bud-producing plants should not develop buds and roots under favorable conditions. It is always worth while to try them in the case of new and rare plants of which stock is wanted quickly. Leaves to be used as cuttings should not be too young. Those that are full grown are to be preferred; such will generally be found in the middle part of the shoot. Some recommend that the petiole be inserted for its whole length; others cut it off. The leaf should be inserted, up to the base of the blade, in pure white sand, laid over sandy peat, or other compost suitable for the growth of the plant. Or the leaf should be laid flat and the base slightly inserted in the sand, in which it should be kept by a small peg or stone and covered with a bell-glass, the edges of which ought to be well pressed into the sand. The glasses must be shaded, and the atmosphere kept moist.

Buds are formed on the principal nerves as well as on the midrib. It is often a good plan to cut these nerves across in several places without, however, severing the lamina. Thus treated a single leaf is capable of producing a large number of plants. Generally callus is formed on the lower part of the severed portion, and from this roots are emitted before any leaf-bud is perceptible. Plants raised in this way assume the character of seedlings. They always, however, reproduce the characters of the parent plant. Buds are, in some cases, emitted from the indentation of the leaf-margin, as in Bryophyllum proliferum.

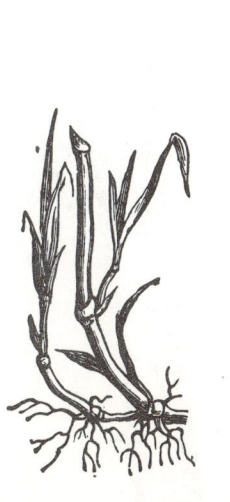

MODE OF LAYERING CARNATIONS

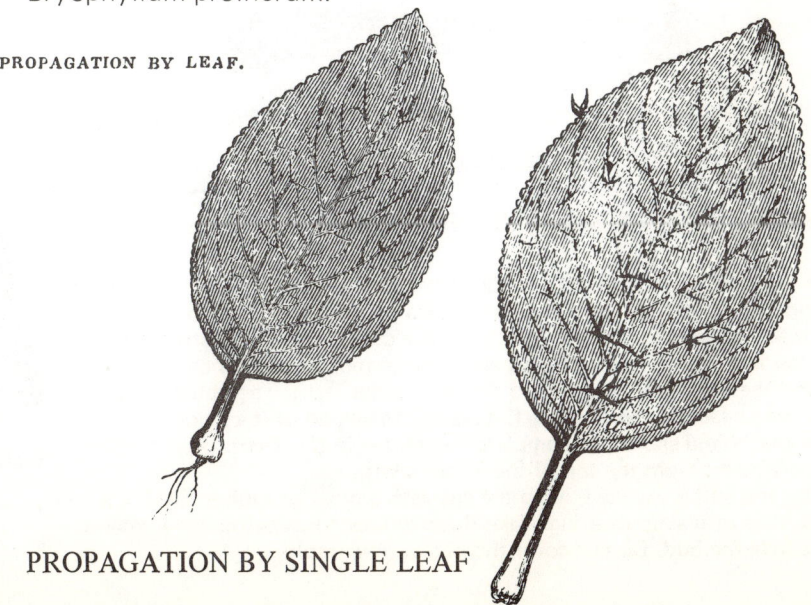

PROPAGATION BY LEAF.

PROPAGATION BY SINGLE LEAF

GRAFTS

For grafts to be successful, the transferred part—the scion—must be from a plant that is closely related to the rooted part—the stock. For example: Graft pink dogwood scion on white dogwood stock, or graft hybrid tea rose scion on multiflora rose stock.

The growing tissues of the scion and stock must be in close contact for the two parts to unite. This growing tissue—the cambium—is the soft layer of cells between the bark and the wood.

Many types of grafts are used for the propagation of ornamental trees and shrubs. The most useful types for the home propagator are bud grafts and cleft grafts.

Bud Grafts

Bud grafts are useful for propagating a large number of plants from a small amount of scion material; only one leaf bud is needed to form each new plant.

Make bud grafts any time during the growing season when the bark of the stock will peel easily from the wood and dormant buds are available.

For stocks, use seedlings or rooted cuttings. A good size of stock for budding is 3/16 to 3/8 inch in diameter—about the thickness of a pencil.

Cut bud sticks from the desired variety. The buds should be plump but dormant. Cut off the leaf about one-fourth of an inch from the bud. The piece of leaf stem that is left protects the bud and is useful as a handle for holding the bud.

Using a very sharp knife, make a T-shaped cut in the bark of the stock. Begin the stem of the T near the ground line and cut upward about 1 inch. Then make the crosscut at the top of the vertical cut. The crosscut should extend about one-third of the way around the stock. When making the T, cut only through the bark, not into the wood.

Use the point of your knife to lift the bark along both sides of the vertical cut.

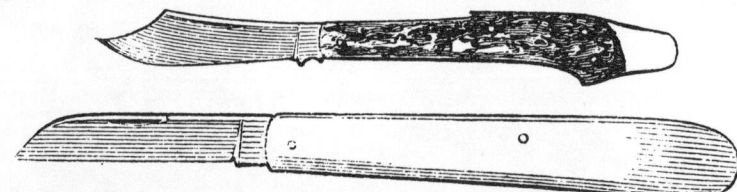

COMMON AND FRENCH BUDDING-KNIVES

Now cut the bud. Start the cut about one-fourth of an inch below the bud. Cut under the bud only deep enough to take a thin sliver of wood. After the knife blade passes beneath the bud, angle the cut upward and outward to remove the bud with a shield of bark about three-fourths of an inch long.

Insert the lower part of the bud shield into the T-cut. Then push it down so the cut surface of the shield is flat against the wood of the stock.

The bud shield should be completely enclosed in the T-cut. If part of the shield protrudes from the top of the T, cut it off.

After the bud is inserted, wrap the cut with a piece of rubber band or a narrow strip of plastic sheeting. Take three or four turns below the bud and again above the bud. Do not cover the bud with wrapping.

Three to five weeks later, cut the wrapping away; the bud should be united with the stock in this time.

Buds usually remain dormant until the next season. In early spring, cut off the top of the stock plant at a point about one-half inch above the bud. This will force the bud to sprout; all growth from the bud will be similar to the bud-source plant.

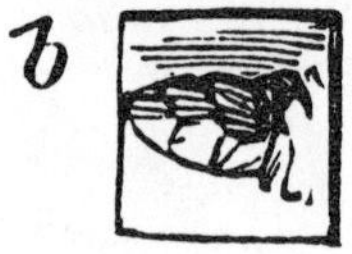

Square Shield-budding

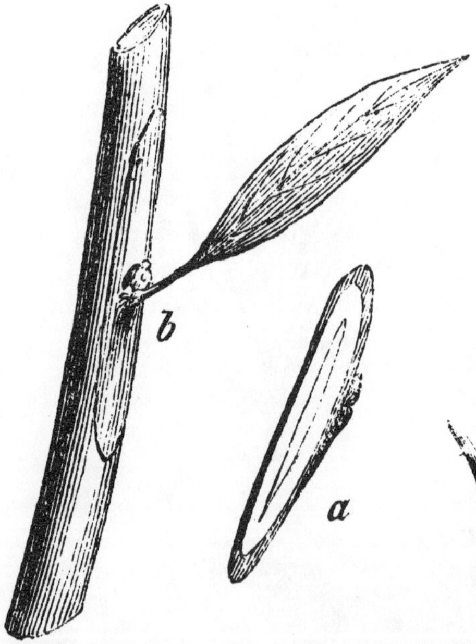

SHIELD-BUDDING THE CAMELLIA

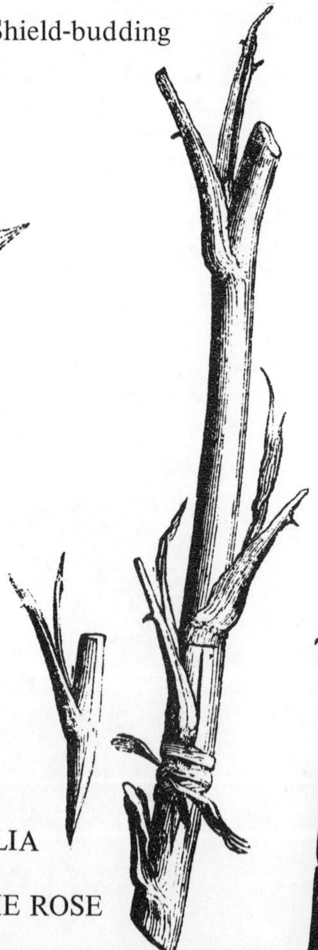

SHIELD-BUDDING THE ROSE

SHIELD-BUDDING REVERSED

Budding a Rose

Budding roses is somewhat more complicated than taking hardwood cuttings but using the proper technique it is possible. Most of the commercial rose growers still bud roses. Pick a nice healthy bud and cut as indicated in the top of the drawing. The larger stock then receives the bud and the wound is bound. A budding knife is a must.

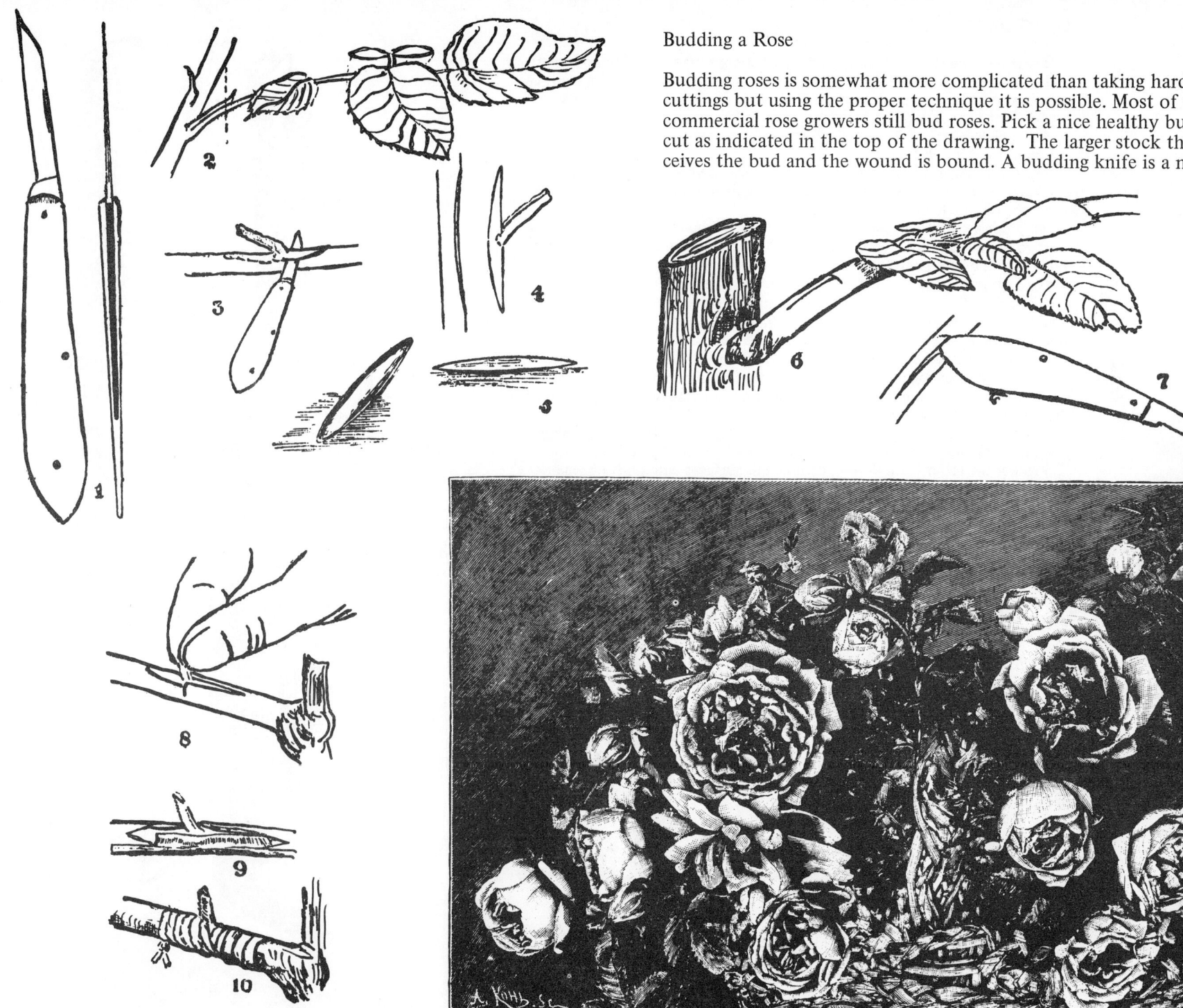

Cleft Grafts

Large new plants can be propagated in a short time if scion wood is cleft grafted to the main stem of the stock plant. Or several varieties of a plant can be grown on the same main stem if scion wood is cleft grafted on heavy branches of the stock plant.

Make cleft grafts while the stock and scion plants are dormant. Late winter is a good time for cleft grafting.

For scion wood, use material from the previous year's growth. For each stock, cut two scions, each of which includes three buds. Cut the scions about 1 inch below the lowest bud.

To prepare the stock, saw it off squarely at the point where you wish the graft to be. With a broad chisel or a stout knife, split the end of the stock to a depth of 2 or 3 inches. Place a wedge in the split to hold it open.

Now prepare the scions. With a sharp knife, carefully trim the butt of each scion to the shape of a wedge. Begin the cuts on each side of the lowest bud. Make the wood on the bud side of the wedge a little thicker than the wood on the opposite side.

Insert the scions in the split stock, with the lowest bud to the outside. The cambium layers of the stock and scions should be in contact. If the scions are set in the stock at a slight angle, the cambium layers are more likely to meet than if the scions are set straight.

When the scions are in place, remove the wedge from the stock. Pressure from the split stock should be sufficient to hold the scions tightly.

Protect the scion from drying; coat all cut surfaces with grafting wax or tree-coating compound—both of which are available from garden-supply stores—or cover the graft with a polythene bag tied around the stock.

If you use a polythene bag, shade the graft to protect it from overheating; direct sunshine can kill the scion. Leave the bag on the graft until new growth fills the bag—probably in late spring. Then remove the bag and cover all cut surfaces with tree-coating compound.

At the end of the first growing season, inspect the scions and cut off the weaker of the two. Cover the stub of the weaker scion with tree-coating compound.

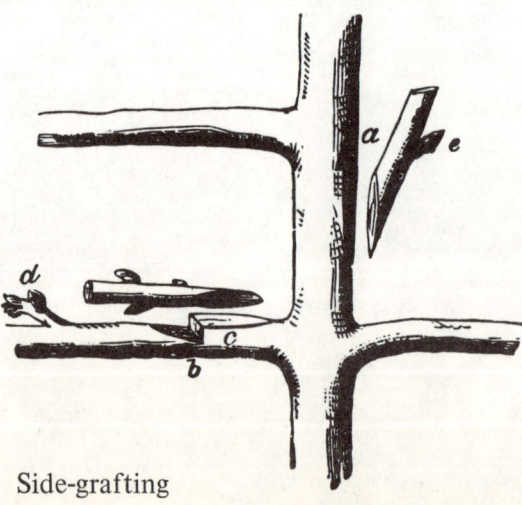

Side-grafting

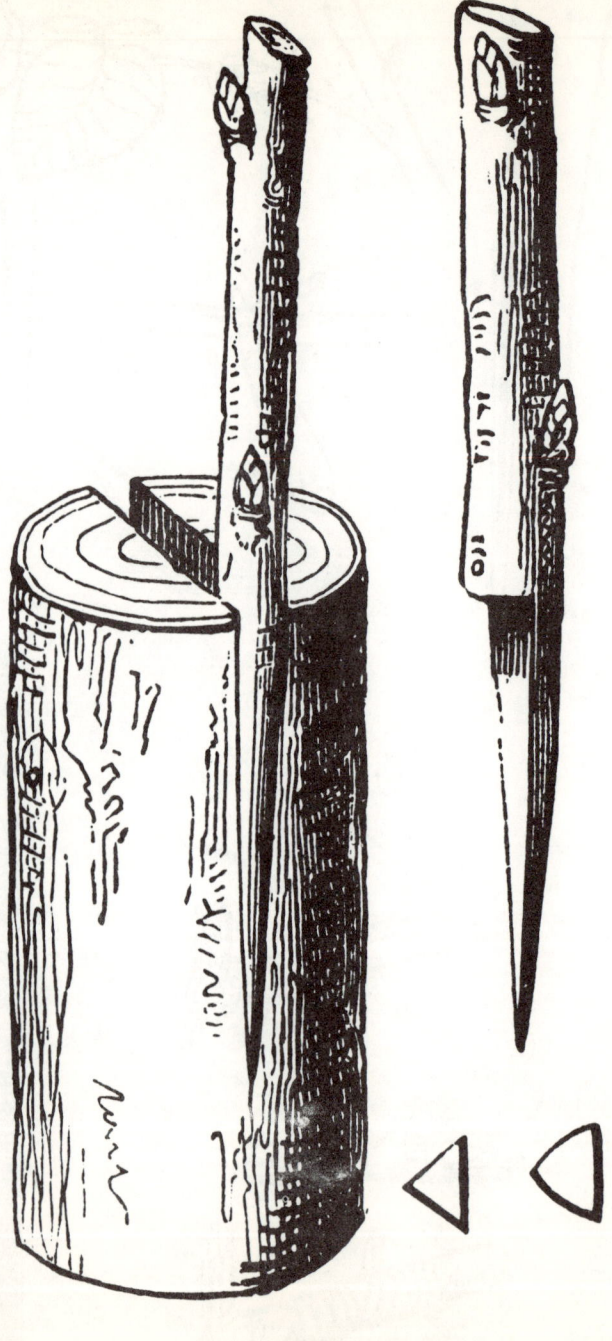

Cleft-grafting

PROPAGATION BY INARCHING

This mode of propagation, also called grafting by approach, is based on the same principles as grafting; in the latter, however, a shoot is entirely detached from a plant, and made to grow upon another; whle in inarching, scion as well as stock is nourished by its own roots until a union is formed.

Inarching was formerly employed for uniting two or more trees for picturesque effect. In rustic gardens, for instance, doorways were formed by planting two trees of the same kind, one on each side of the intended entrance; these were trained upright to the desired height, and then their tops were bent to unite, so as to form but one head. Trees to form arbors, etc., may be so united, or the stems of several trees may be inarched to a central one, which may ultimately be rendered independent of its own stem and roots.

There are various ways of inarching, the simplest being where A is the stock, B the plant to be inarched upon it. The two may be growing in the ground, or one may be growing in a pot and the other in the ground, or both may be in pots. At a convenient place where A and B can be brought in contact, as between a and b, cut off corresponding slices from each; then bind the parts together, and clay or otherwise protect, as in grafting. The stock may be allowed to remain at full length, or it may be cut back to c or d, and afterwards to a. When the two have formed a union, B may be separated from its own roots, by cutting it off in the direction of b, thus leaving it wholly dependent for support upon the roots of the plant a. Before this final separation is made, it is advisable to wean off gradually the portion be from its original source of nourishment, by making an incision below b, deepening it from time to time, till at last there is but little communication left between B and its proper roots; then that little may be cut off without causing any material difference to the inarched part B/E. In stead of diminishing the connection between the inarched part and its own roots, by gradually cutting in at or below b, it is a good plan to take off a narrow ring of bark, when the nature of the plant will permit. This may be done by degrees as the union is effected. If buds are retained on A, with the view of maintaining the necessary amount of circulation in the stock till that can be done by B, care should be taken to check the shoots that push from them; for it allowed to grow vigorously, they would attract the sap from the part B. On referring to the figure, it will be observed that on the separation being effected at b, there must be a heel left at that place, which will take some considerable time to disappear.

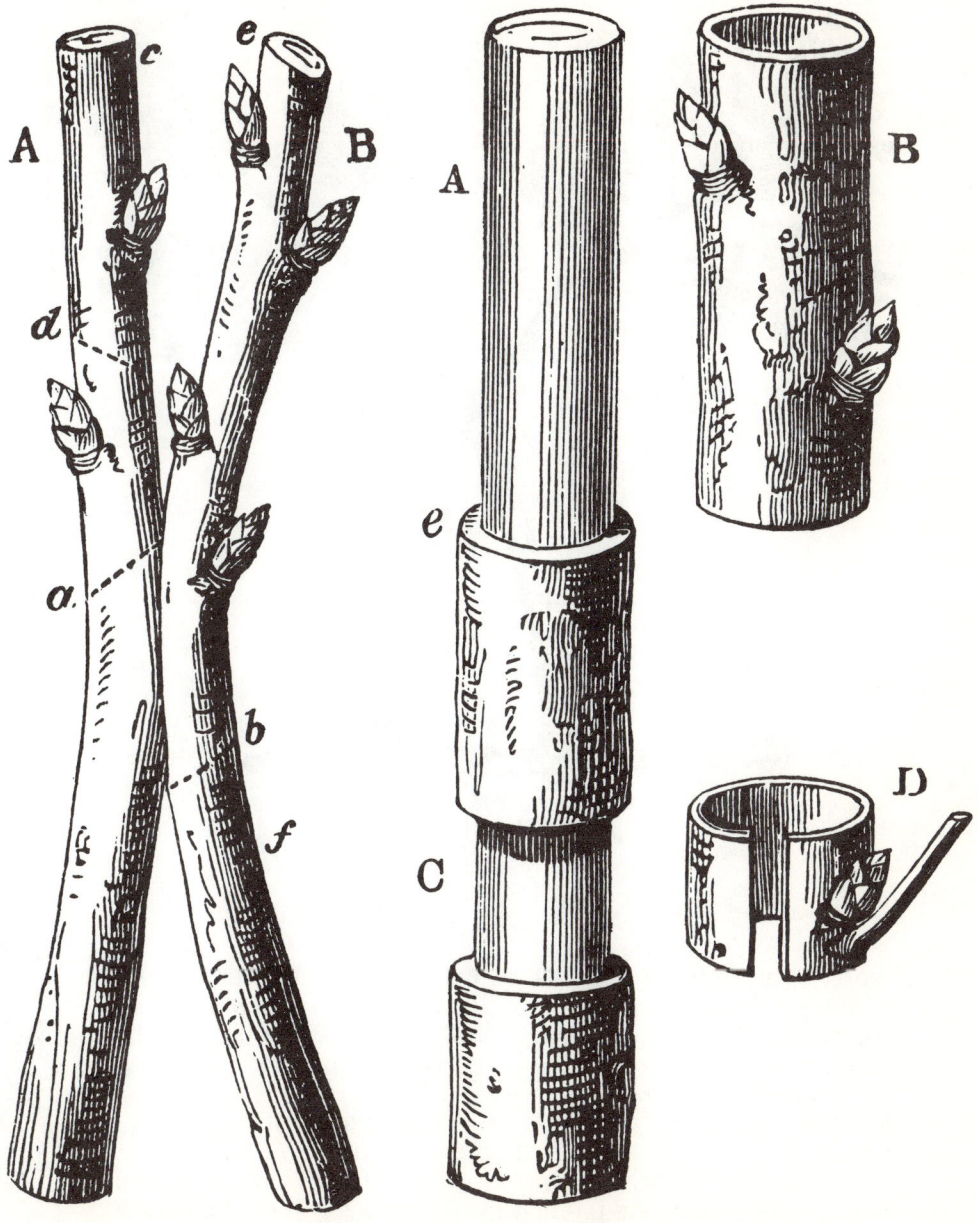

Inarching

Flute and Ring budding

Sylvan Inarching

The stems of young trees may be inarched to form trelliswork, or an arch. Summer shoots may also be inarched on shoots of the same age, or on a stem or branch several years older than themselves. In this way branches which have died or become diseased may be readily replaced by others. The parts to be united should be firmly bound and held securely to prevent friction by wind. If the bark is removed from the parts to be placed in contact, the union will be all the quicker. Quaint arbors have been constructed by planting a number of young trees in a circle, then bending them over and binding them firmly together at a height of about 12 feet. Ultimately they unite and appear to be one tree.

Two ash trees naturally grafted together

HOW TO TRIM COLEUS

Cut misshapen or overgrown plants back at "A". Pot resulting cutting

A. When seedlings have developed two true leaves (left), transplant them to another container. The plants on the right have been allowed to get too large before transplanting.

TRANSPLANTING

B. Using a knife blade, carefully lift seedlings from the planting container.

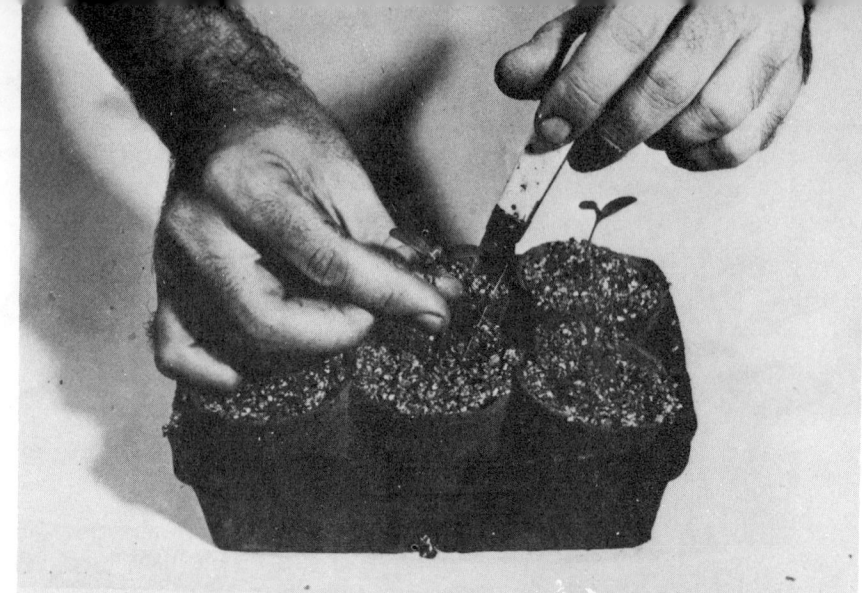

C. Make a slit with the knife blade in the vermiculite in the new container and set the seedling in the slit. Firm the vermiculite around the roots with your forefingers, taking care not to crush the seedling.

D. Fertilize seedlings twice a week. Seedlings on the left were not fertilized; those on the right, the same age, were fertilized twice weekly.

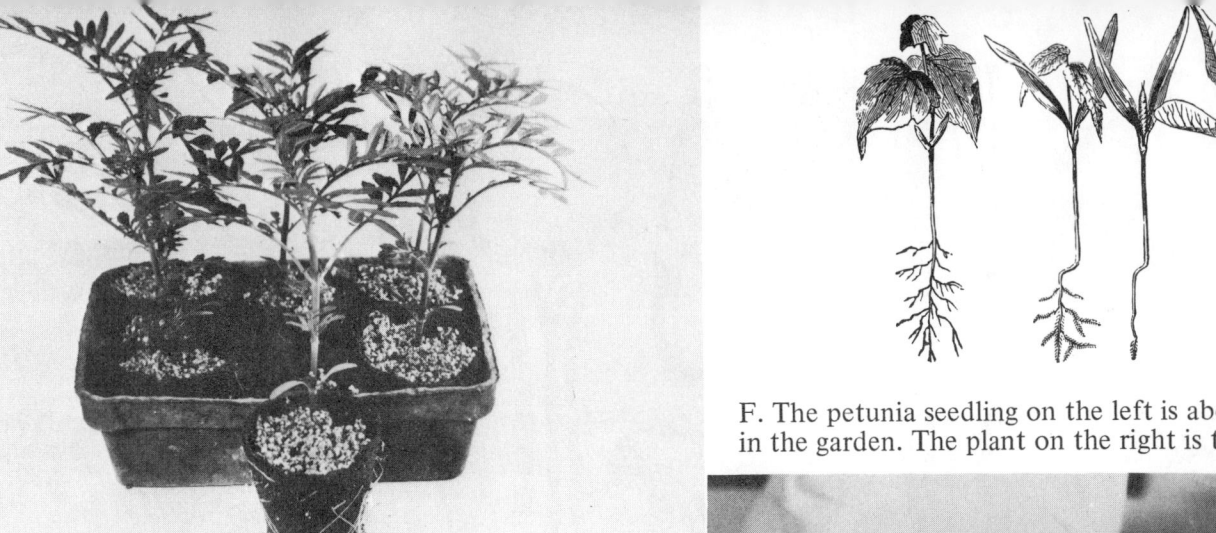

Germination of the Maple

F. The petunia seedling on the left is about the right size for setting in the garden. The plant on the right is too large.

E. Seedlings, 8 weeks old, ready to be transplanted to the garden. Note roots growing through the walls of the peat pot.

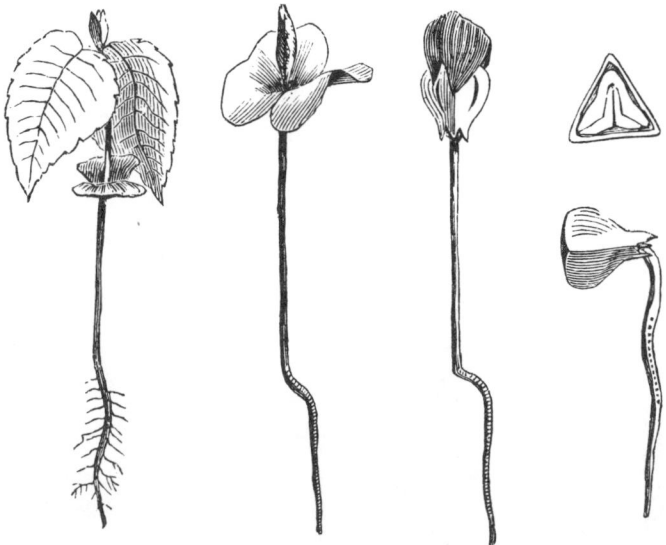

Germination of the Beechnut

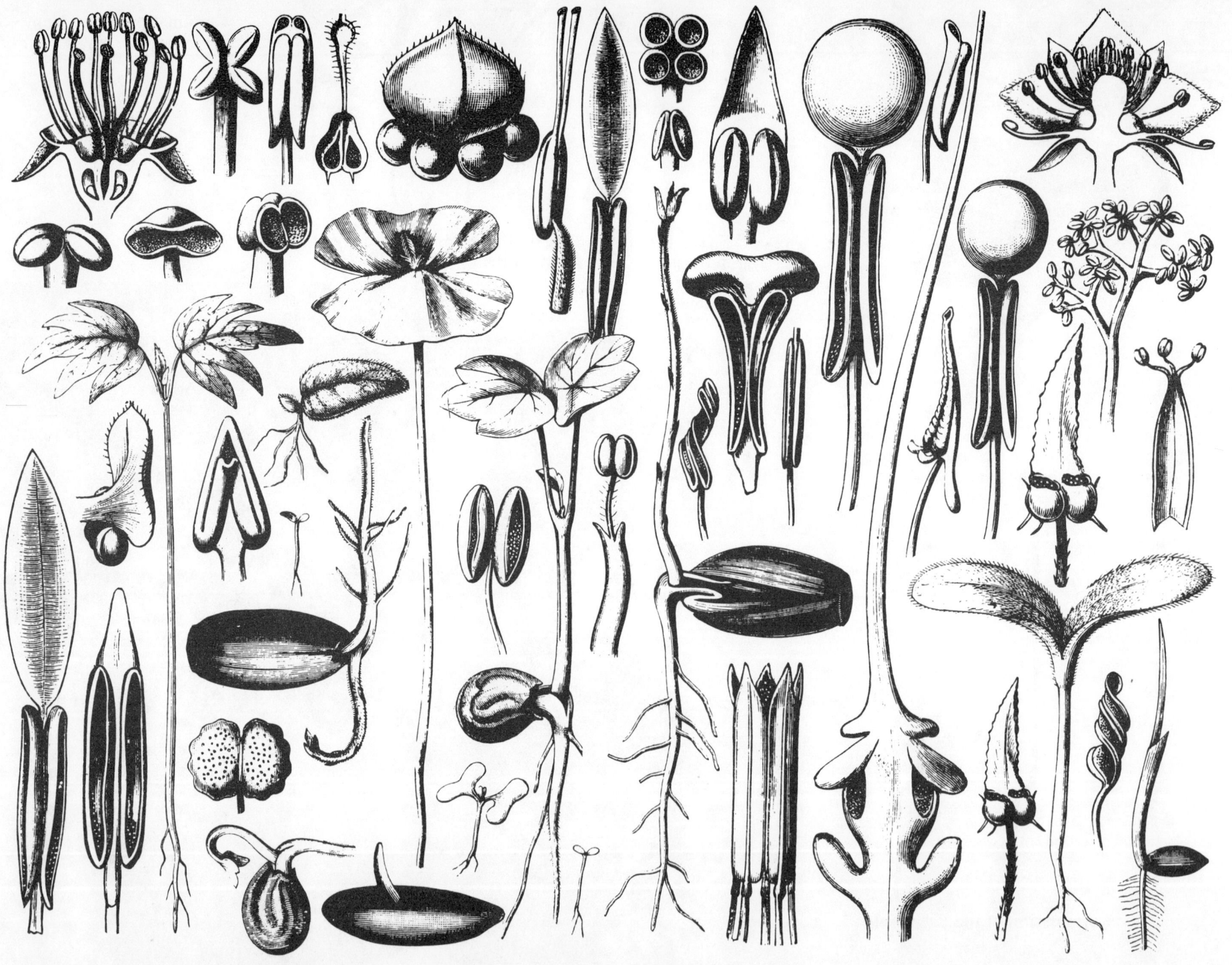

PLANT SEXUAL SYSTEMS

The reproduction of plants falls into two categories. The first is the vegetative or asexual reproduction. This includes all the natural methods of propagation that do not involve sexual activity. New plants that develop by runners, suckers, tubers, corms, rhizomes and root suckers are included in this vegetative process. The mechanical or human-controlled propagation such as grafting and bulbing are covered in the propagation section. Many of the natural vegetative means of propagation also are used in controlled methods in the garden. We can reproduce by means of layers and suckers but they do occur naturally without the help of the gardener.

The most common method of reproduction in the plant world is seed. This method involves sexual activity. The drawing on this page demonstrates all the sexual parts of the flowers and the next few pages will show the complex activity of fertilization and pollination and the beginnings of plant life.

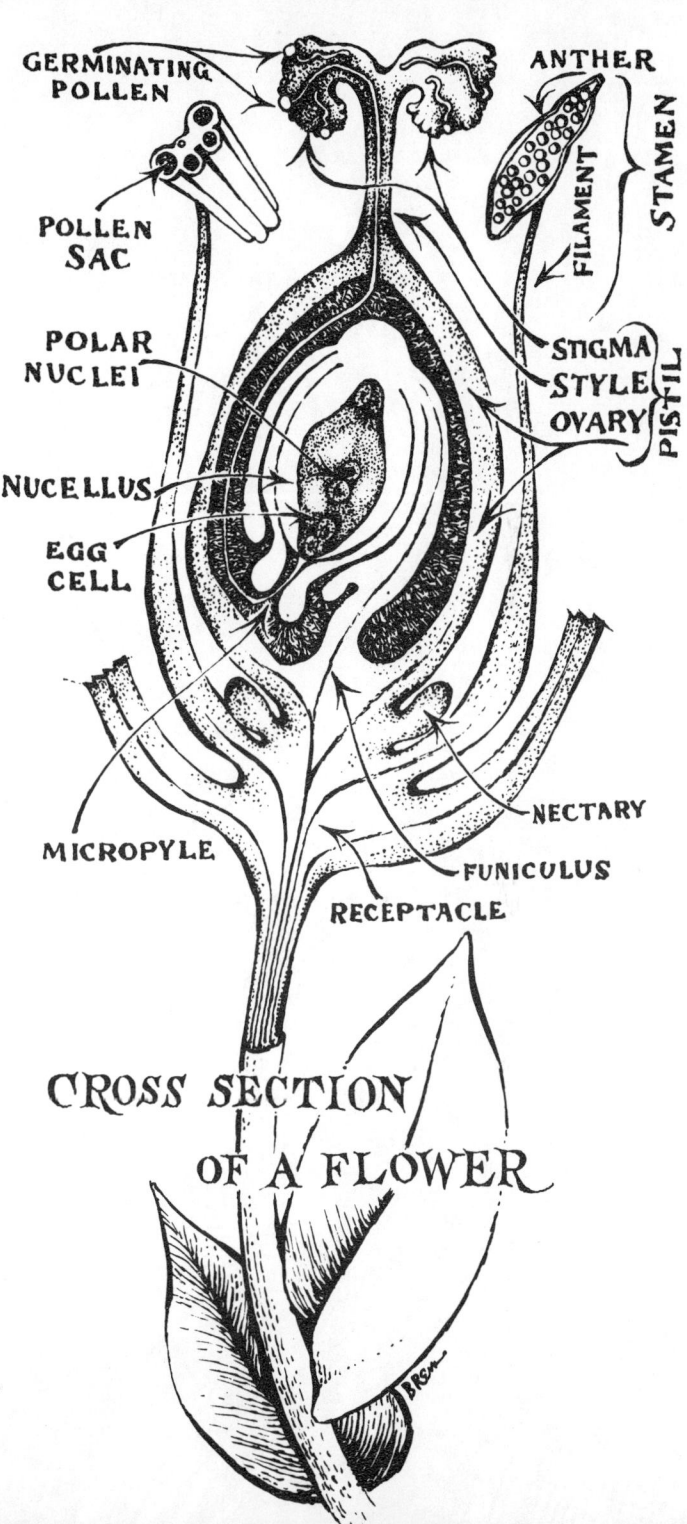

CROSS SECTION OF A FLOWER

Gardeners interested in simple botany will find this Analytical Arrangements of Botanical Terms valuable in recalling the terms used in the description of any parts of the plant. For the plant collectors or botany students collecting plant parts for further study the guide should prove an invaluable reference.

ORDER OF DESCRIPTION.
(Specimen Page.)

Bluets.
Houstonia caerulea.

Root	Multiple, fibrous.
Stem	Herb, erect, branching, 4′, glabrous.
Leaves	Sessile, indistinct, oblong to spatulate, acute, entire, glabrous, opposite, 3″ to 5″ long.
Flowers	Solitary, terminal, slender peduncle, perfect, 5″.
Calyx	Green, cleft, half-inferior cup.
Sepals	4, pointed, valvate.
Corolla	Light-blue with yellow eye, cleft, perigynous, salver.
Petals	4, oval, pointed, valvate.
Stamens	4, sessile, free, epipetalous, alternate.
Anthers	Oblong, 2, longitudinal
Pistil	1, compound, style 1.
Stigma	2, oblong, lateral.
Ovary	2, half-superior, axillary.
Fruit	Loculicidal, pod notched at apex.
Seeds	Several in each cell, saucer-shaped.

Remarks. — In some plants the style is exserted and the stamens included, in others the style is included and the stamens exserted. They grow in tufts.

Class. — Exogenous.

Division. — Monopetalous.

Order. — Rubiaceæ.

Name. { Scientific. — Houstonia cærulea.
{ Common. — Common Houstonia or Bluets.

Locality. — Moist meadows, Trenton, N. J.

Analytical Arrangements of Botanical Terms

ROOTS.

Kinds. 1. PRIMARY, growing from root-end of embryo.

a. SIMPLE. Conical, 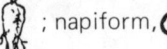 ; napiform, ; fusiform, .

b. MULTIPLE. Moniliform, 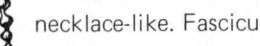 necklace-like. Fascicu-

lated, tufted, thick and fleshy. Tubercular,

 having small tubers. Fibrous,  thread-

like.
2. SECONDARY, growing from stems.
Underground, starting from stem below ground. Aerial, starting from stem above ground.

Conical Root

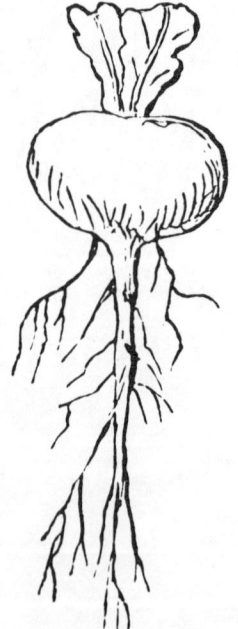
Napiform Root

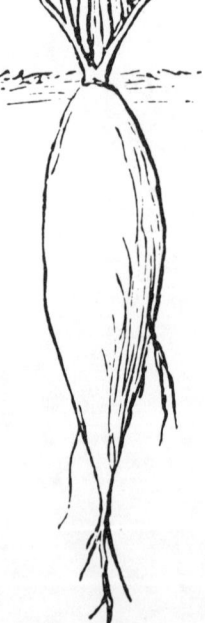
Fusiform Root

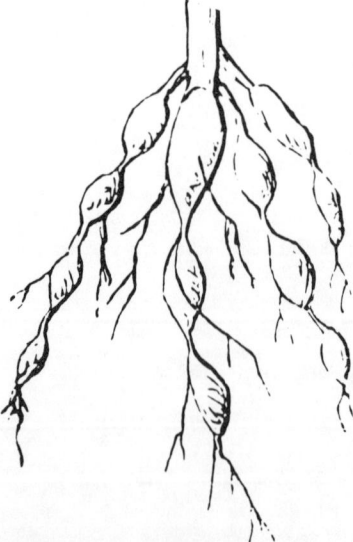
Moniliform Root

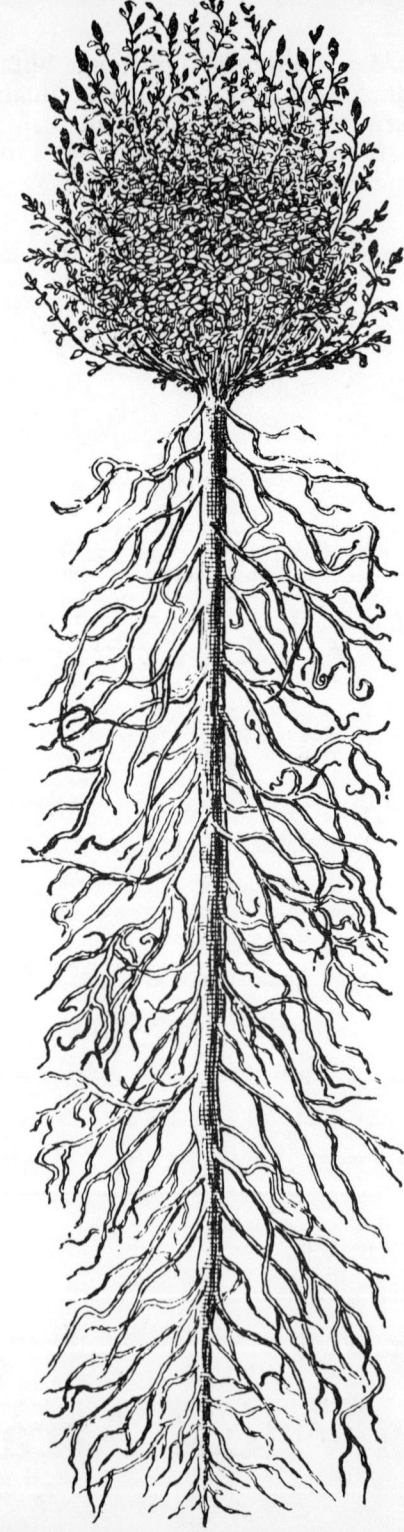
Fasciculated Root

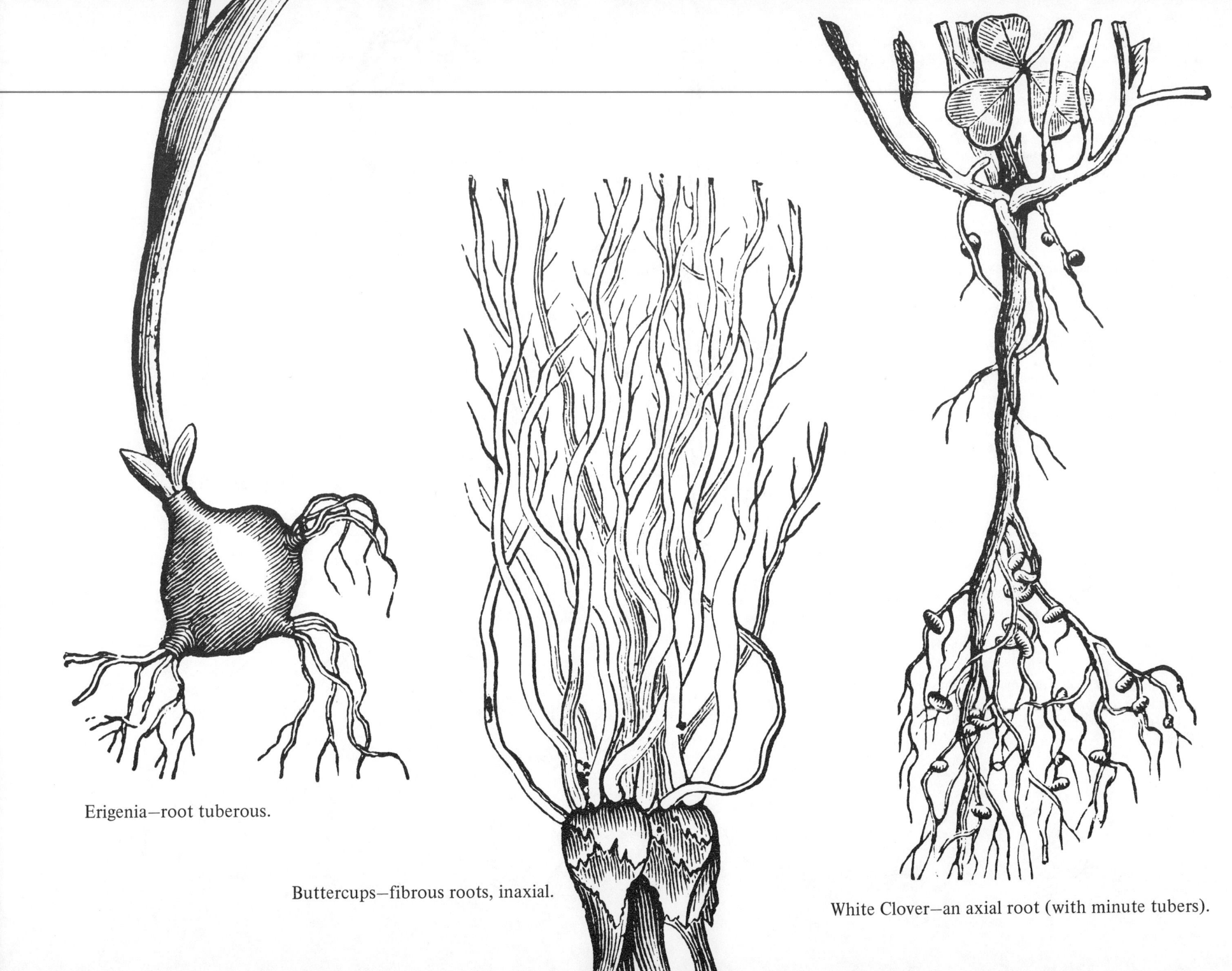

Erigenia—root tuberous.

Buttercups—fibrous roots, inaxial.

White Clover—an axial root (with minute tubers).

STEM.

Parts. n, Node, part to which the leaf is fastened.
i, Internode, portion between nodes.
a, Axil, the angle between leaf and stem, upper side.

Class. Exogenous, outside-growing (Maple, Elm).
Endogenous, inside-growing (Corn-stalk, Timothy).

Situation. 1. Above ground, usually leaf-bearing.
2. Under ground, scale-bearing.

Stems above Ground.

Character. Herbaceous, soft, not woody (Four-o'clock).
Suffrutescent, slightly shrubby (Toad-flax).
Suffruticous, shrubby at base (Trailing Arbutus).
Fruticous, shrubby (Currant-bushes).
Arborescent, tree-like (Flowering Dogwood).
Arboreous, tree (Elm).

Direction of Growth. Repent, prostrate and rooting from the under surface (Partridge-berry).
Procumbent, prostrate, but not rooting (Purslane).

Decumbent, prostrate, except at the extremity (Poor Man's Weather-glass).

Assurgent, ascending obliquely.

Erect, upright (Indian Corn).

Scandent, climbing with tendrils or rootlets (Grape, English Ivy).

Voluble, twining (Morning-glory).

Declinate, declined or bent downwards (Blackberry).

Diffuse, loosely-spreading (Red Currant).

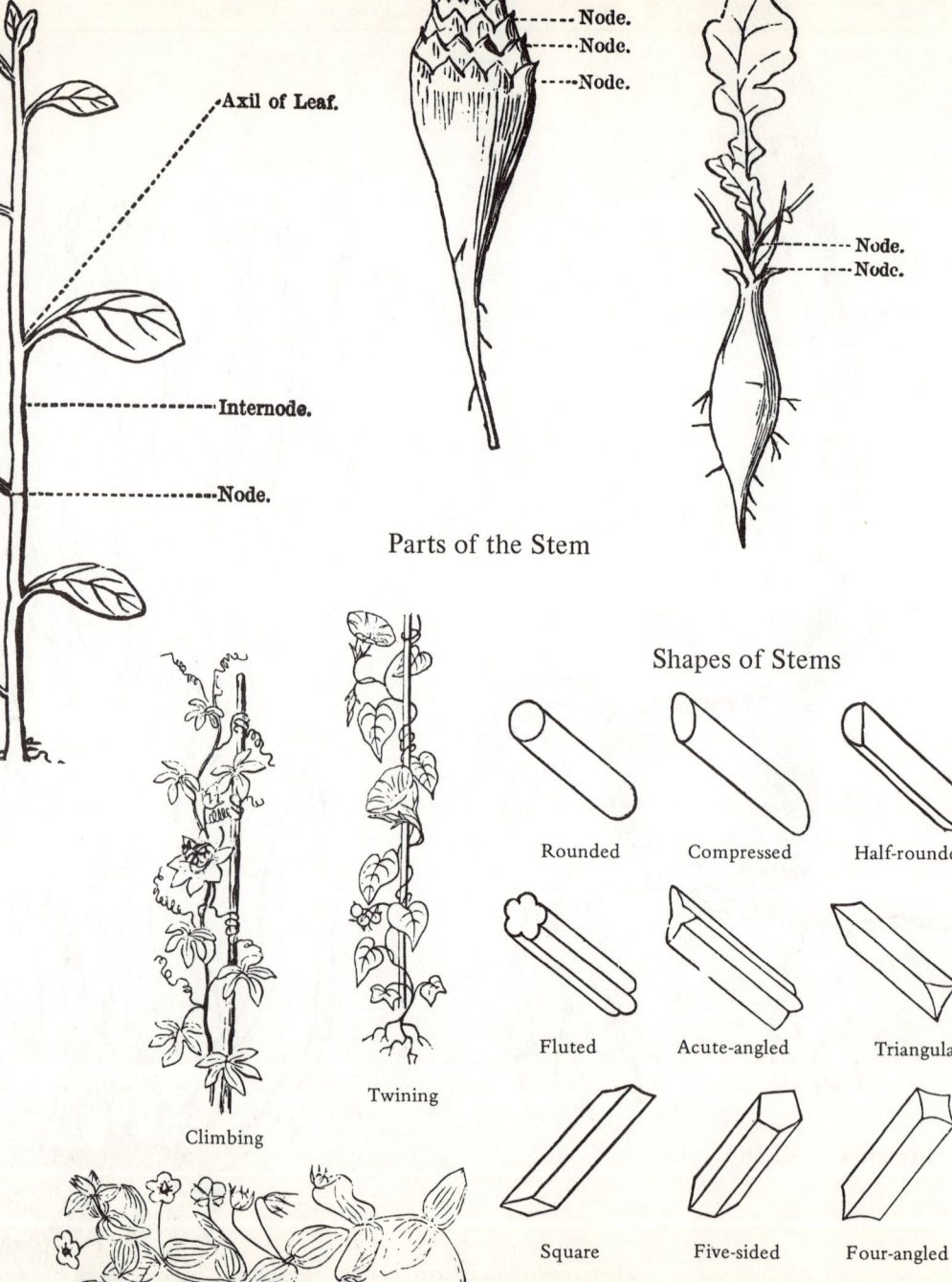

Parts of the Stem

Shapes of Stems

Ascending   Creeping   Climbing   Twining   Trailing

Rounded   Compressed   Half-rounded
Fluted   Acute-angled   Triangular
Square   Five-sided   Four-angled

Forms of Branches. Sucker,  a branch of subterranean origin that finally rises out of the ground. The Raspberry multiplies in this way.

Offset,  a short, prostrate-rooting branch with a tuft of leaves at the end (Houseleek).

Runner,  a long, prostrate-rooting branch with tuft of leaves at the end (Houseleek).

Stolon,  a branch that curves downward and takes root. The Currant multiplies in this way.

Tendril,  a thread-like coiling branch used for climbing.

Spine or Thorn,  a hard, sharp-pointed branch.

Stems under Ground.

Kinds. Rhizoma or Rootstock,  a perennial, horizontal stem, partially or wholly subterranean (Calamus).

Tuber,  an enlarged stem with eyes (White-potato).

Bulb,  a bud, usually subterranean with flesh scales (Onion, Lily).

Corm,  a solid bulb (Indian Turnip).

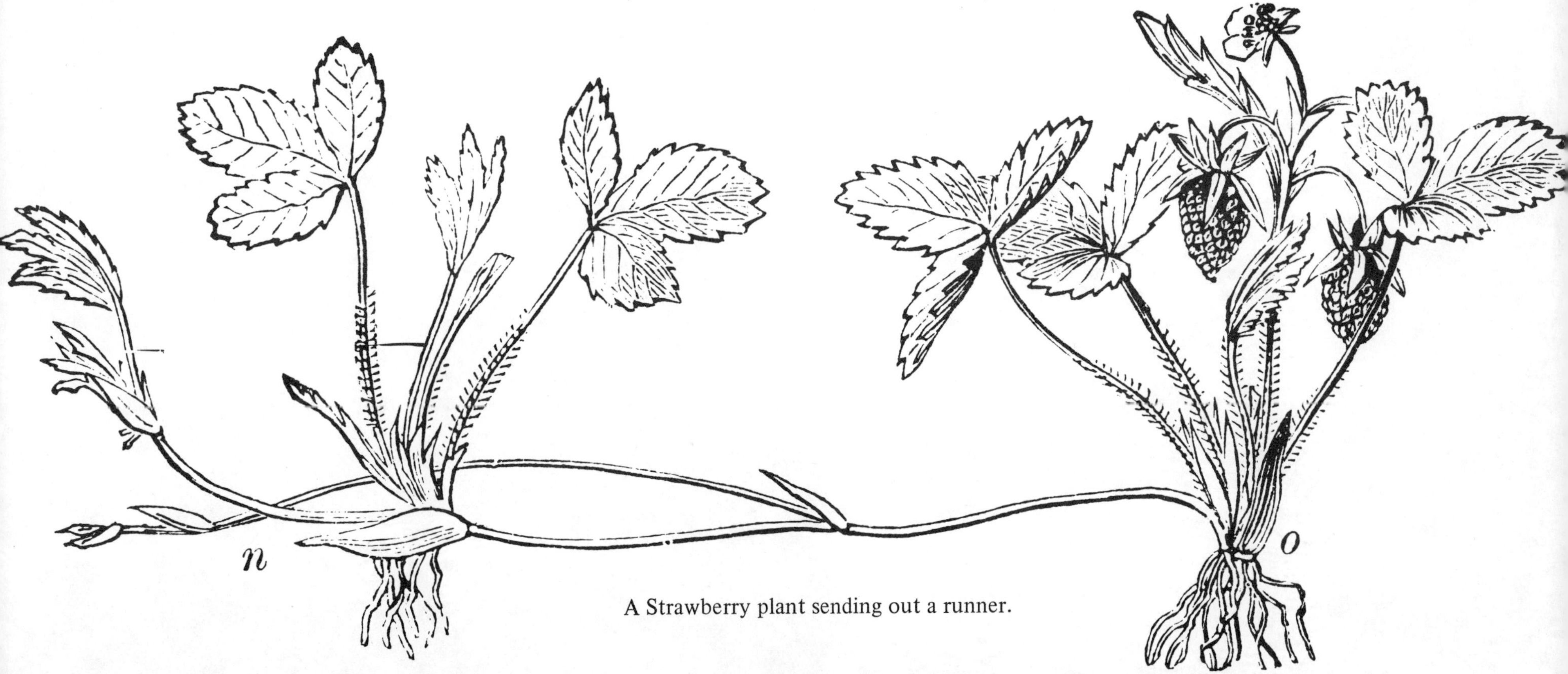

A Strawberry plant sending out a runner.

LEAVES.

Parts. b, Blade, the expanded portion.
p, Petiole, the stem.
s, Stipules, leaf-like appendages at base of petiole.

Kinds. 1. SIMPLE, having but one blade.

Sessile, without petiole

Petiolate, with petiole.

Stipulate, with stipules.

Cirrhous, with tendril.

2. COMPOUND, having more than one blade.

a. Pinnate, with leaflets arranged along a common petiole.

Abruptly pinnate, with even number of leaflets.

Odd-pinnate having an odd leaflet

Unipinnate, divided but once.

Bipinnate, divided twice.

Tripinnate, divided three times.

b. Palmate, leaflets diverging from one point.

Unipalmate, divided but once.

Bipalmate, divided twice.

Tripalmate, divided three times.

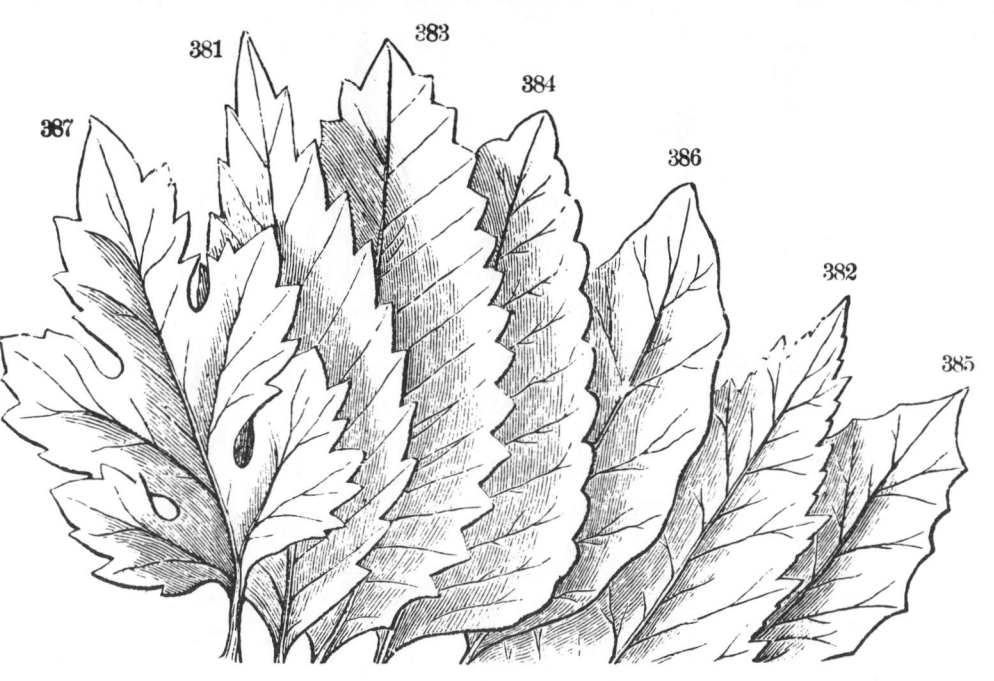

Serrate leaf of Chestnut. 382, Doubly serrate leaf of Elm. 383, Dentate leaf of Arrow-wood. 384, Crenate leaf of Catmint. 385, Repand leaf of Circaea. 386, Undulate leaf of Shingle Oak. 387, Lobed leaf of Chrysanthemum.

COMPOUND LEAVES

Three-fingered   Five-fingered

Adnate Stipules

Varieties of venation.—1, Feather-veined,—leaf of Betula populifolia (White Birch), lying upon a leaf of Plum-tree; same venation with different outlines. 2, Palmate-veined,—leaf of White Maple, contrasted with leaf of Cercis Canadensis. 3, Parallel venation,—plant of "three-leaved Solomon's seal" (Smilacina trifoliata). 4, Forked venation,—Climbing Fern (Lygodium).

Framework. Midrib, the central vein.

Ribs, strong veins branching from near the base of midrib.

Veins, the branching framework.

Veinlets, small veins.

Venation. Parallel, with simple veins running parallel from base to apex.

Feather, with lateral veins branching at regular intervals from midrib.

Radiate, with strong veins branching from apex of petiole.

Reticulate, with veins and veinlets that unite and separate in the form of network.

Form. a. BROADEST AT THE MIDDLE. Peltate, ; orbicular, ; oval, ; elliptical, ; oblong, ; linear, ; acerose, (Pine).

b. BROADEST AT BASE. Deltoid, ; ovate ; lanceolate, ; subulate, ; cordate, ; reniform, ; hastate, ; sagittate,

c. BROADEST AT THE APEX. Obovate, ; oblanceolate, ; spatulate, ; cuneate, ; obcordate, ; lyrate, ; runcinate,

Bases. Auriculate, ; oblique, ; tapering, ; abrupt, ; clasping, ; perfoliate, ; connate, ; decurrent,

Apexes. Obcordate, ; emarginate, ; retuse, ; truncate, ; obtuse, ; acute, ; acuminate, ; mucronate, ; cuspidate, ; aristate,

Margins. Entire, ; repand, ; sinuate, ; crenate, ; dentate, ; serrate, ; incised, ; laciniate, ; palmately-lobed, ; palmately-cleft, ; palmately-parted, ; palmately-divided, ; pinnately-lobed, ; pinnately-cleft, ; pinnately-parted, ; pinnately-divided,

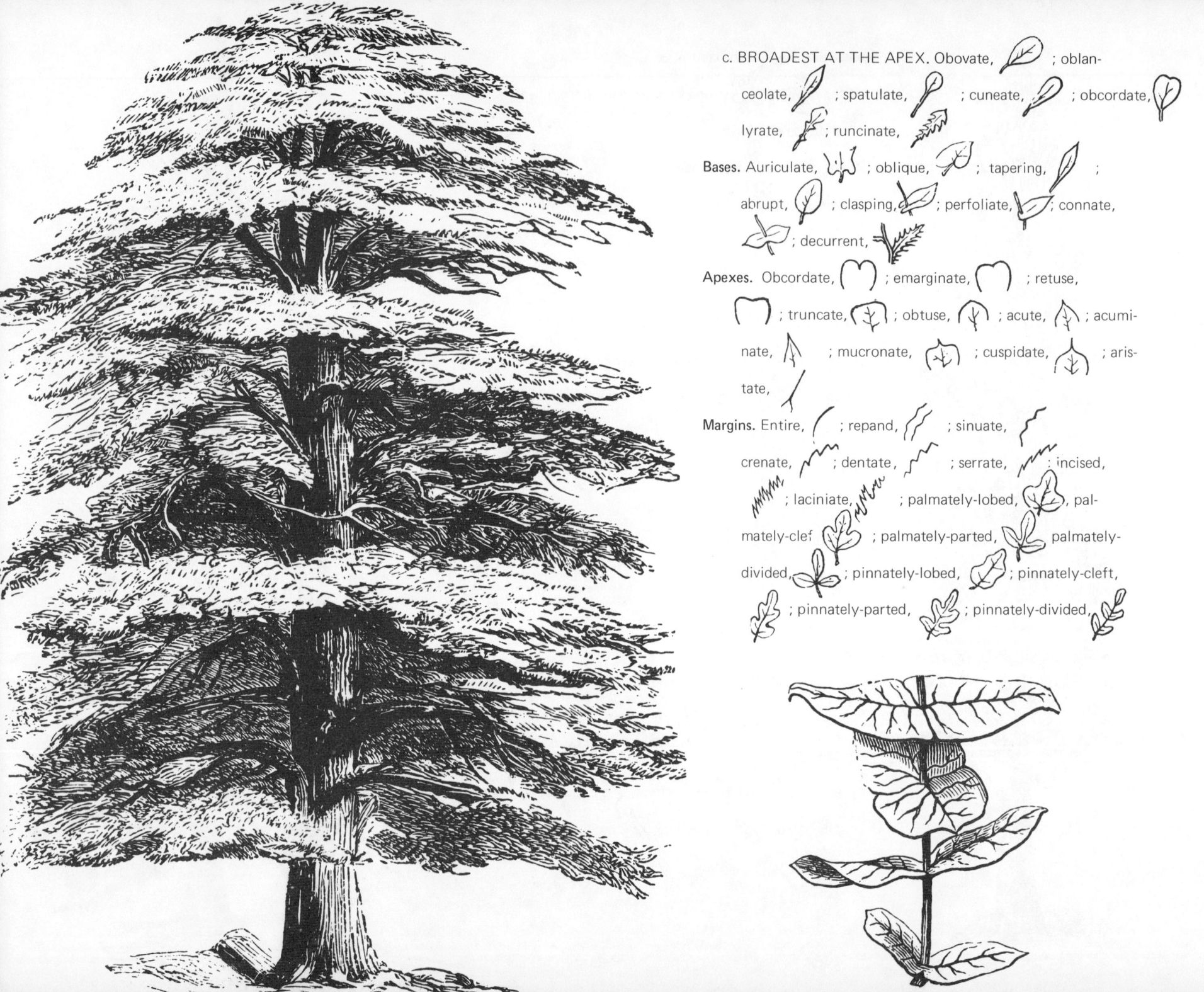

Surface. a. WITHOUT HAIRS. Glabrous, smooth.
   b. SOFT HAIRS. Pilous, few, short; hirsute, few, long; pubescent, dense, short; villous, dense, long; sericeous, silky; lanuginous, woolly; tomentous, matted like felt; floccous, fleecy tufts.
   c. STIFF HAIRS. Scabrous, minute, hard points; hispid, few, short points; setous, bristly; spinous, having spines.

Color. Glaucous, covered with whitish powder.
   Canescent, grayish-white with fine pubescence.
   Incanous, hoary-white.
   Punctate, having transparent dots.
   Hyaline, nearly transparent.

Texture. Succulent, fleshy; coriaceous, leather-like; scarious, dry; rugous, wrinkled.

Phyllotaxis, arrangement on the stem. Alternate, ; opposite, ; whorled (verticillate); radical near the ground; cauline, on the stem; rosulate, clustered; fasiculate, in bundles.

Vernation, arrangement in the bud.

   Induplicate, folded crosswise (Tulip-tree).

   Conduplicate, folded along midrib (Oak).

   Plicate, folded like a fan (Red-currant).

   Circinate, rolled lengthwise (Fern).

   Convolute, rolled edgewise (Cherry).

   Involute, both edges rolled inward (Apple).

   Revolute, both edges rolled outward (Willow).

   Equitant, astraddle (Iris).

   Obvolute, half equitant (Jerusalem Sage).

   Triquetrous, triangular equitant (Sedges).

Duration. Fugacious, falling very early.
   Deciduous, falling at the close of the season.
   Persistent, remaining through the winter.

INFLORESCENCE.

Parts. Flower,  the blossom.

Peduncle, 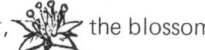 the stem of a solitary flower or the main stem of a flower-cluster.

Scape, a peduncle that grows from the ground.

Pedicel,  ; p, the stem of each flower of a flower-cluster.

Bracts, b, small floral leaves.

Involucre, a cluster of bracts.

Catkin   Raceme

Clustered Terminal

Stem Leaves

ATTITUDE OF INFLORESCENCE

Erect   Nodding   Radical Leaves

Lady's slipper (leaves alternate); 2, Synandra grandiflora (leaves opposite); 4, Medeola Virginica (leaves verticillate); 3, Larix Americana (leaves fasciculate).

Kinds. 1. SOLITARY, single, alone.
  Terminal, at the summit of the stem.

  Axillary, 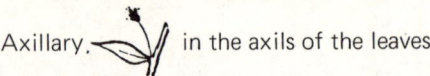 in the axils of the leaves.

2. CLUSTERED, several flowers collected in a bunch.
  a. INDEFINITE or INDETERMINATE, flowering from axillary buds. Inflorescence centripetal.

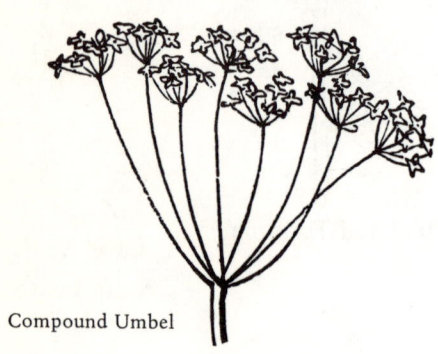

Compound Umbel

  Raceme,  flowers arranged along the axis; pedicels about equal in length (Currant).

  Corymb, same as raceme, except that the lower pedicels are elongated, making the top flat (Hawthorn).

  Umbel, same as corymb, except that the pedicels branch from about the same point (Milkweed).

  Panicle, compound raceme (Blue-grass).

  Thyrsus, a compact panicle (Lilac).

  Spike, same as raceme with flowers sessile (Mullein).

  Spadix, a fleshy spike, generally enveloped by a large bract called a Spathe, (Calla Lily).

FLOWERS PEDICELLATE.

Solitary Terminal

FLOWERS SESSILE

  Ament or Catkin, 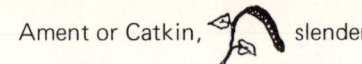 slender pendent spike, with scaly bracts (Birch).

  Head or Capitulum, a shortened spike, reduced to a globular form (Clover).

b. DEFINITE or DETERMINATE, flowers all terminal. Inflorescence centrifugal.

  Cyme, flat-topped or rounded inflorescence (Elder).

  Fascicle, a compact cyme (Sweet-William).
  Glomerule, a cyme condensed into a head (Mint).

  Verticillaster, two opposite glomerules joined (Motherwort).

  Scorpioid, a one-sided and coiled cyme (Forget-me-not).

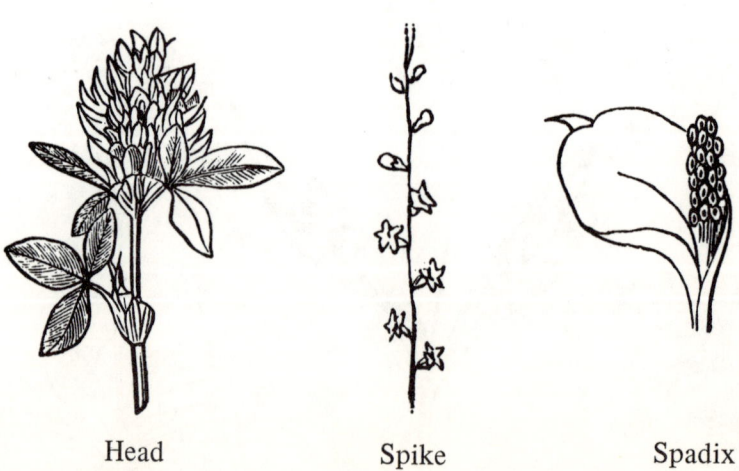

Head   Spike   Spadix

SIMPLE

FLOWER

Parts. Receptacle, the part upon which the several organs of the flower are inserted.

Calyx,  The exterior floral envelope.

Corolla,  the interior floral envelope. The calyx and corolla constitute the protecting organs, sometimes called perianth.

Stamens,  the fertilizing organs.

Pistils, The seed-bearing organs. The stamens and pistils constitute the essential organs.

Kinds. Symmetrical.  same number in each set of organs; unsymmetrical, different number.

Complete,  all the sets present; incomplete, some sets wanting.

Regular  sepals and petals uniform; irregular,  sepals or petals unlike.

Perfect, stamens and pistils both present; imperfect, one set absent.

Staminate, with stamens only; pistillate, with pistils only; neutral, with neither.

Monoecious, staminate and pistillate on same plant; dioecious, on different plants.

Dichlamydeous, having calyx and corolla; monochlamydeous, having calyx only; achlamydeous, having neither.

Di,  tri, tetra, penta-merous,  two, three, four, or five parts in each set.

Sessile, without peduncle; pedunculate,  with peduncle.

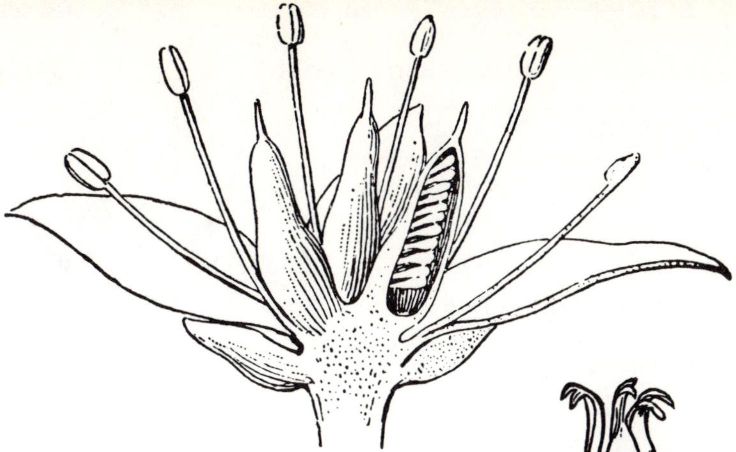

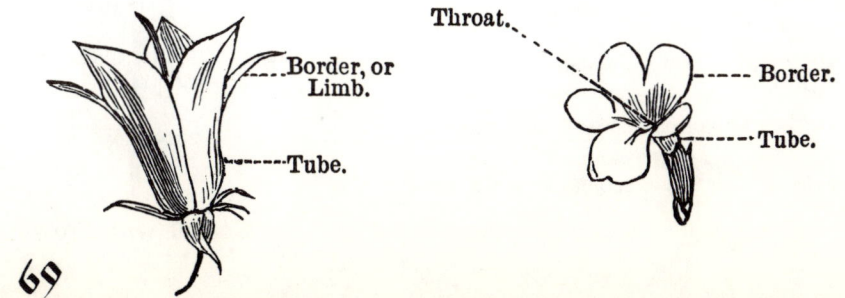

Flower of stonecrop

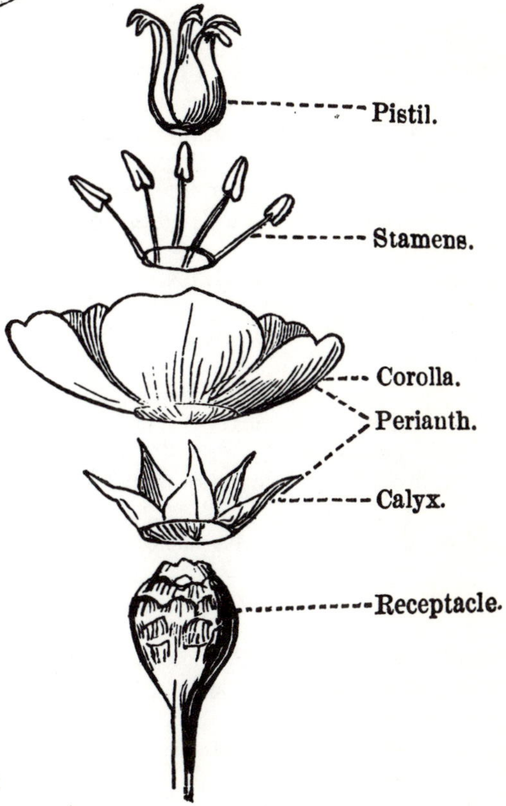

Parts of a Gamopetalous Corolla

DEVIATIONS FROM THE NORMAL OR PATTERN FLOWER ARISE FROM

Augmentation, increase of floral circles (Water Lily).
Chorisis, increase of organs by division. The Bleeding-heart shows the collateral chorisis of stamens, and the Catchfly show the transverse chorisis of corolla.

Anteposition, parts opposite instead of alternate (Grape).

Cohesion, 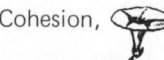 union of parts of the same set (corolla of Morning-glory).

Adnation, union of different sets. In the Cherry the stamens and corolla are inserted upon the calyx.

Irregularity, parts of the same set unequally developed (Violet, Pea).

Suppression, non-development of some parts. In the mints some of the stamens are suppressed or wanting.

CALYX.

Parts.—Sepals, the division of the calyx

Tube, the united portion of a gamosepalous calyx.
Teeth or lobes, the distinct or divided portion of a gamo-sepalous calyx.
Throat, the orifice or summit of the tube.

Pappus, in Compositae, the calyx border consisting of scales, teeth, bristles, or slender hairs.

Cohesion.—Gamosepalous or Monosepalous, 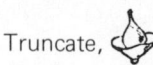 sepals partially or wholly grown together.

Truncate, 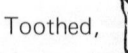 without lobes.

Toothed,  lobes small.

Lobed, 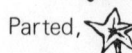 parted about one fourth.

Cleft, parted about one half.

Parted, separated nearly to the base.

Polysepalous,  separated to the base.

Parts of the Calyx.

Calyx

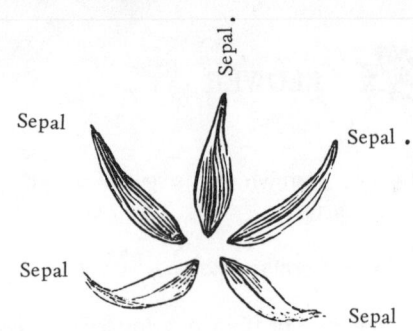
Sepal. Sepal. Sepal. Sepal. Sepal.

SEPAL.—One of the leaves of the calyx.

Adnation.—Inferior, calyx free from ovary.

Half-inferior, calyx adherent to the ovary half-way.

Superior, calyx adherent to the ovary.

Form.—See under COROLLA.
Aestivation.—See under COROLLA.

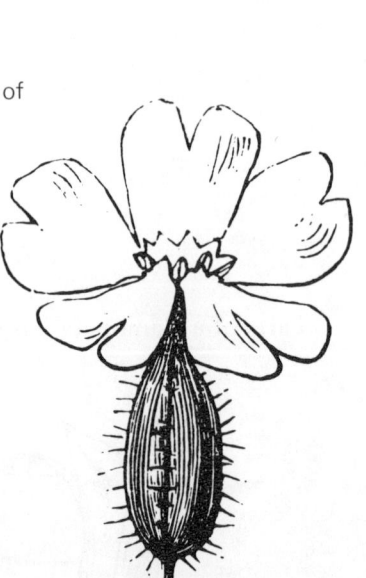
Corolla with Crown

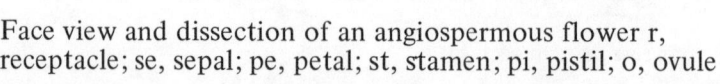

Face view and dissection of an angiospermous flower r, receptacle; se, sepal; pe, petal; st, stamen; pi, pistil; o, ovule

COROLLA.

Parts.—Petals the divisions of the corolla.

Lamina, the expanded portion of the petal.

Claw, the stem portion of the petal.

Spur, 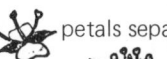 ; s, the hollow portion of certain corollas.

Crown, a small projection from certain petals.

(Catchfly.)

Cohesion.—Gamopetalous or Monopetalous, petals partially or wholly grown together.

Truncate, toothed, lobed, cleft, parted.

Polypetalous, petals separate.

Adnation.—Hypogynous, corolla attached under the pistil (gynia, pistil).

Perigynous, corolla attached to the calyx. It is thus around the pistil.

Epigynous, corolla attached to the ovary. It is thus upon the ovary which is part of the pistil.

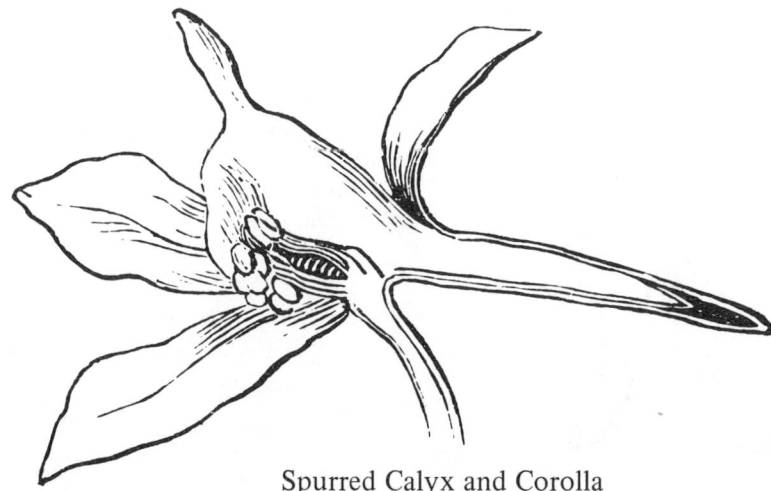

Spurred Calyx and Corolla

KINDS OF COROLLA AND PERIANTH

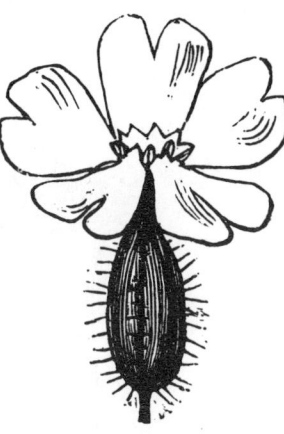

Polypetalous Corolla

Gamopetalous Corolla

KINDS OF REGULAR GAMOPETALOUS COROLLAS

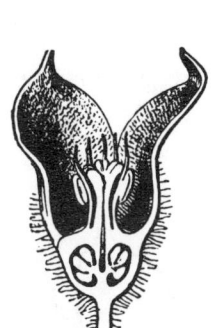

Flower of (European) wild ginger, with calyx but no petals

Tubular

Urceolate

Rotate

  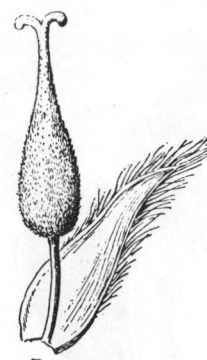

Flowers of willow

A, staminate flower; *B*, pistillate flower. Magnified.—After Decaisne

A   B

Form.—GAMOPETALOUS and POLYPETALOUS.

REGULAR

Urceolate, urn-shaped (Whortle-berry).

Tubular, cylindrical (Trumpet Honeysuckle).

Campanulate, bell-shaped (Hare-bell).

Infundibuler, funnel-shaped (Morning-glory).

Hypocraterimorphous, salver-shaped (Phlox).

Regular Gamopetalous Corolla

GAMOPETALOUS

Rotate, wheel-shaped (Potato).

IRREGULAR

Ligulate, strap-shaped (Dandelion).

Labiate, two-lipped.

Galeate, upper lip arched

Ringent, both lips arched (Dead-nettle).

Personate, throat closed (Toad-flax).

REGULAR

Rosaceous, petals without claws (Rose)

Liliaceous, petals with claws gradually spreading (Lily).

Caryophyllaceous, long claws enclosed in a tube (Pink).

Cruciferous, four clawed petals in the form of a cross (Mustard).

IRREGULAR.

Papilionaceous, butterfly-shaped (Bean).

PARTS.—Vexillum, banner; alae, wings; carina, keel.

Irregular Gamopetalous Corolla

POLYPETALOUS

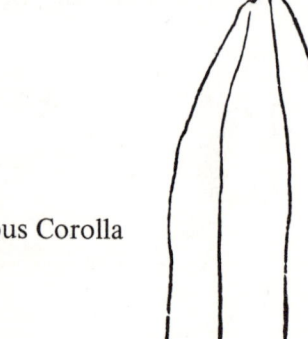
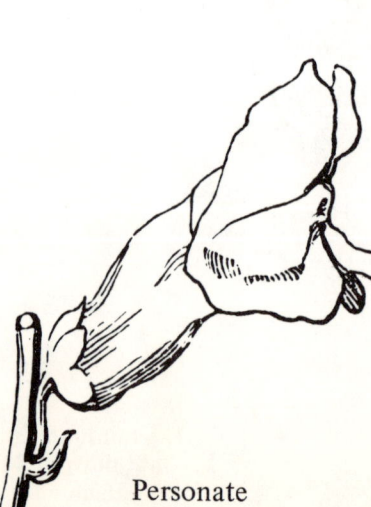

Ligulate   Personate   Ringent

Aestivation, the arrangement of the floral organs in the bud.

Valvular, pieces met by their margins (Lilac).

Induplicate, margins turned inward (sepals of Clematis).

Reduplicate, margins turned outward (sepals of Hollyhock).

Convulute, or contorted, each piece overlaps its neighbor in one direction (Geranium).

Imbricated, one or more petals wholly outside.

Quincuncial, five petals, two without and two within and the remaining one with one edge outside and the other inside.

Triquetrous, three petals, one without and one within, and the remaining one with one edge outside and the other inside.

Vexillary, having one large petal enclosing the others (Pea).

Plicate, the folding of gamopetalous flowers.

Supervolute, with folds turned obliquely in the same direction (Morning-glory).

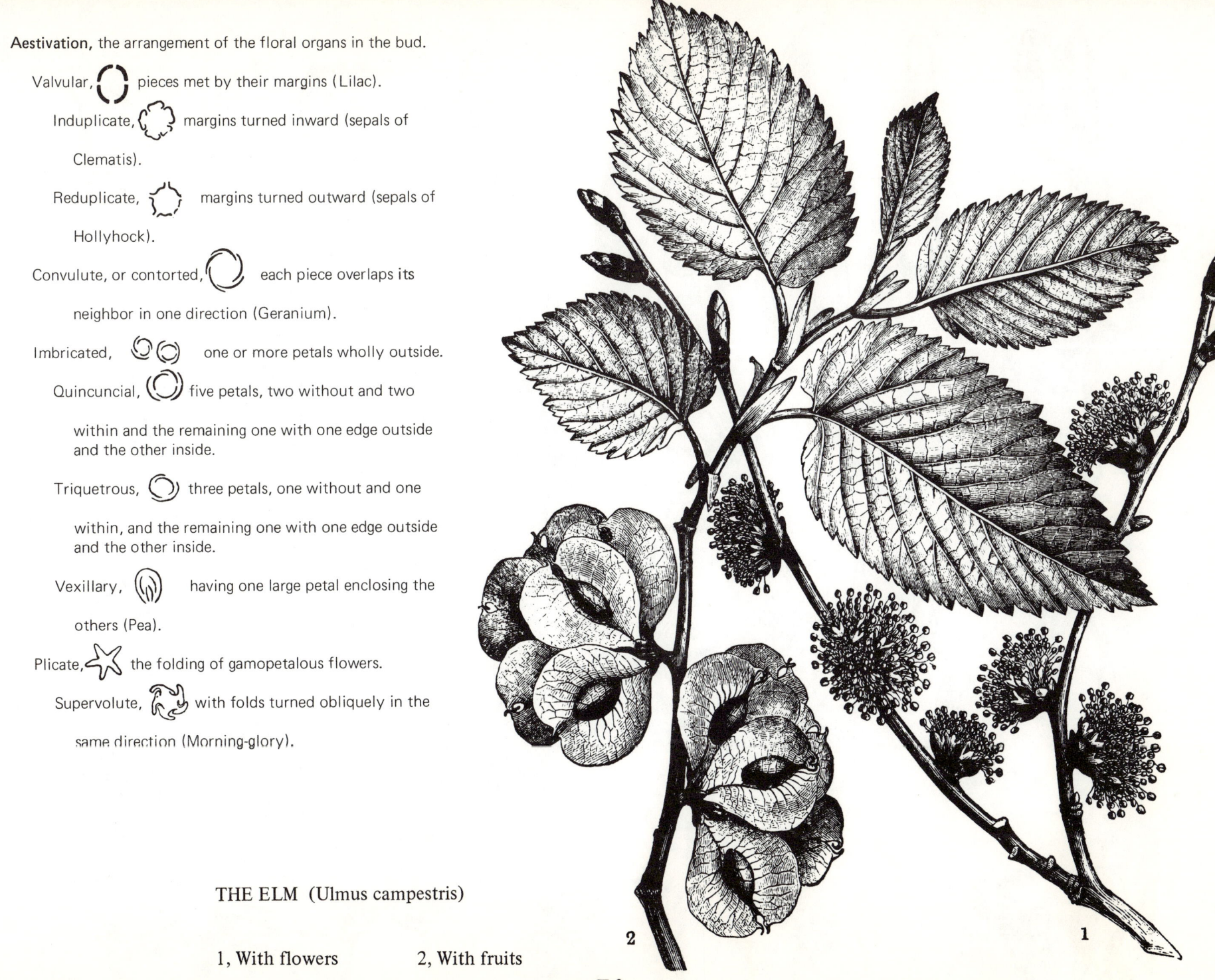

THE ELM (Ulmus campestris)

1, With flowers   2, With fruits

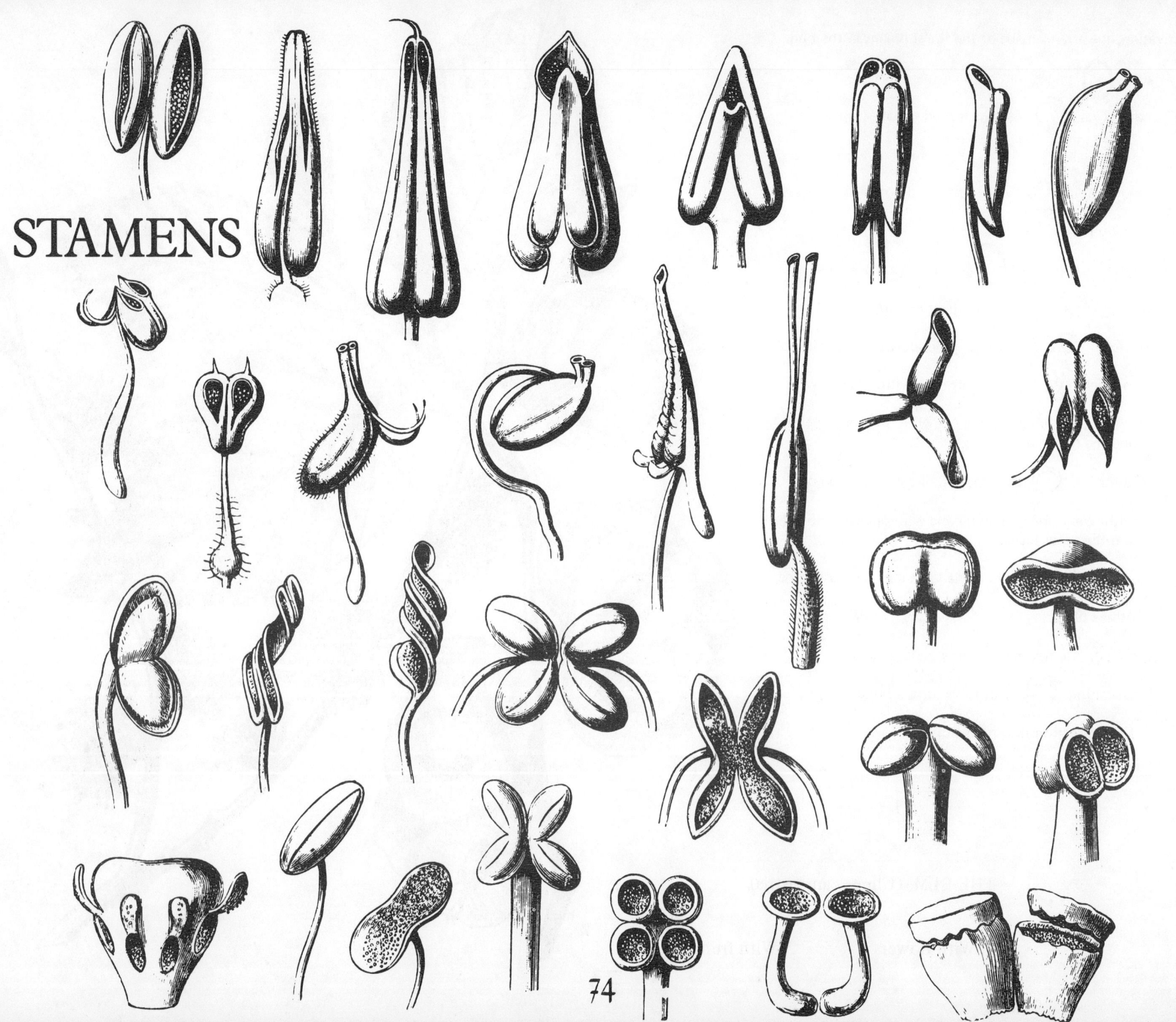

STAMENS (Androecium).

Parts.— Anther, the enlarged and essential portion.

Filament, the stem holding the anther.
Pollen, the fertilizing powder found in the anther.

Kinds.— Sessile, anther without filament.

Sterile, filament without anther.

Connivent, converging

Exserted, protruding out of corolla.

Included, entirely within the corolla.

Didynamous, four in number, two long and two short.

Tetradynamous, six in number, four long and two short.

Cohesion.— Syngensious, united by their anthers.

Monodelphous, united by their filaments into one set.
Diadelphous, united into two sets.
Polyadelphous, united into many sets.

Adnation.— Hypogynous, borne on the receptacle.

Perigynous, borne on the calyx.

Epipetalous, borne on the corolla.

Alternate, alternate with the lobes.

Opposite, in front of the lobes.
Epigynous, borne on the ovary at its summit.
Gynandrous, borne on the style (Orchid).

FILAMENT.

Kinds.— Filiform, subulate, dilated, petaloid, bidentate.

ANTHER.

Parts.— Lobes (thecae) and connective.

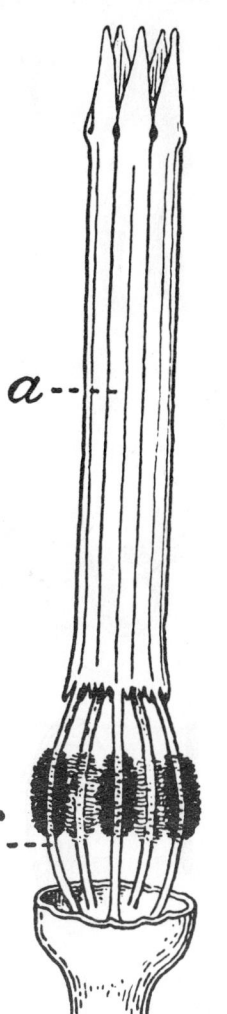

Parts of Stamens.

Stamens of a thistle, with anthers united into a ring

a, united anthers; *f*, filaments, bearded on the sides. — After Baillon

Adnation.— Innate, anther firm on summit of filament.

Adnate, anther attached by its whole length to filament.

Extrorse, facing the petals.
Introrse, facing the pistils.

Versatile, attached near the middle.

Dehiscence.— Longitudinal, opening lengthwise.

Transverse, opening crosswise.

Porous, opening by terminal holes.

Valved, opening by valves or doors.

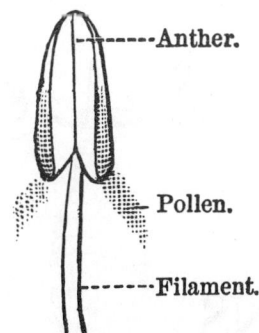

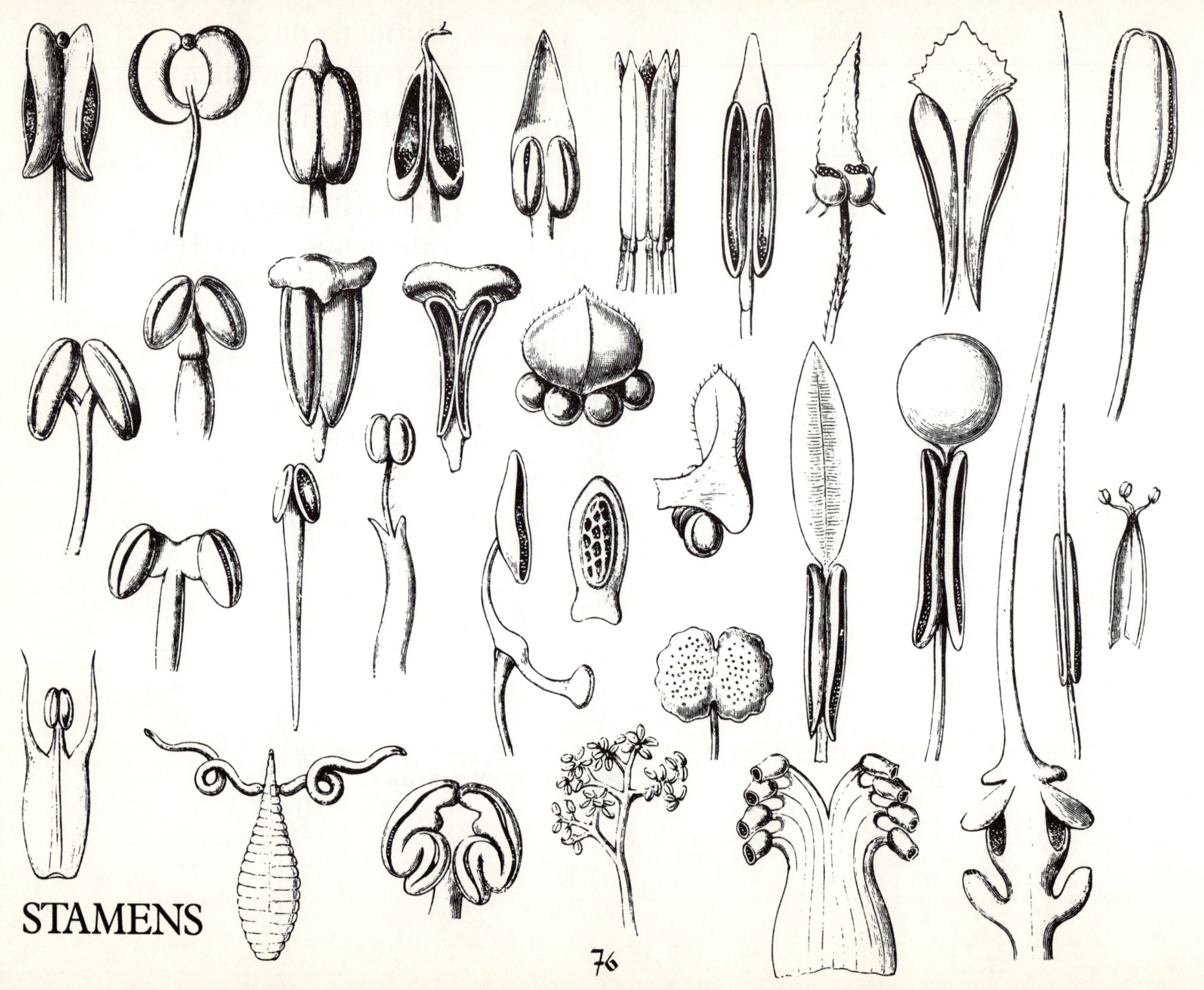
STAMENS

Parts of the Pistil.

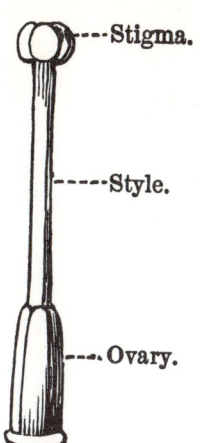

O′VARY.—The lowest part of the pistil, containing the seeds.

STYLE.—The slender stem-like part of the pistil next above the ovary.

STIG′MA.—The top of the pistil.

PISTILS (Gynoecium).

Parts.— Stigma, the rough end to which the pollen adheres.
Style, the stem holding the stigma.
Ovary, the enlarged portion containing the ovules.

Cohesion.—Simple, Having but one cell, placenta style and stigma.

Multiple, a collection of simple pistils (Blackberry).

Compound, simple pistils grown together, each called a carpel.

STIGMA.

Kinds.—Sessile, stigma on ovary; no style.
Globose, globular (Four-o'clock).

Capitate, broad and flat.

Lobed, rounded.
Feathered, like a feather (Grasses).
Linear, thread-like (Corn).

STYLE.

Kinds.—Basal, attached to base of ovary (Forget-me-not).
Lateral, attached to side of ovary (Strawberry).

Terminal, attached to top of ovary.

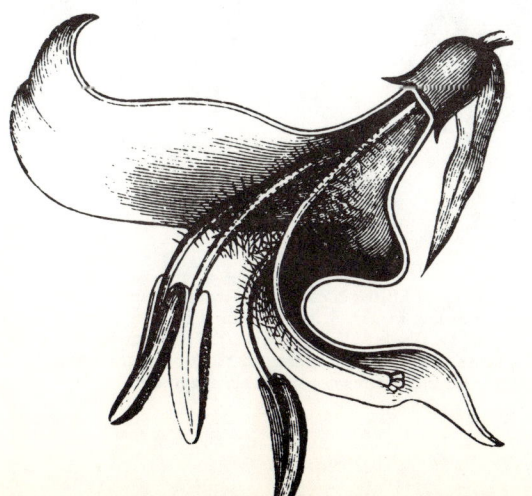

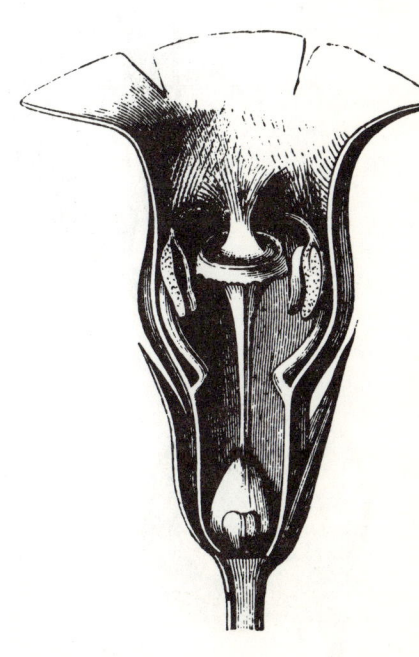

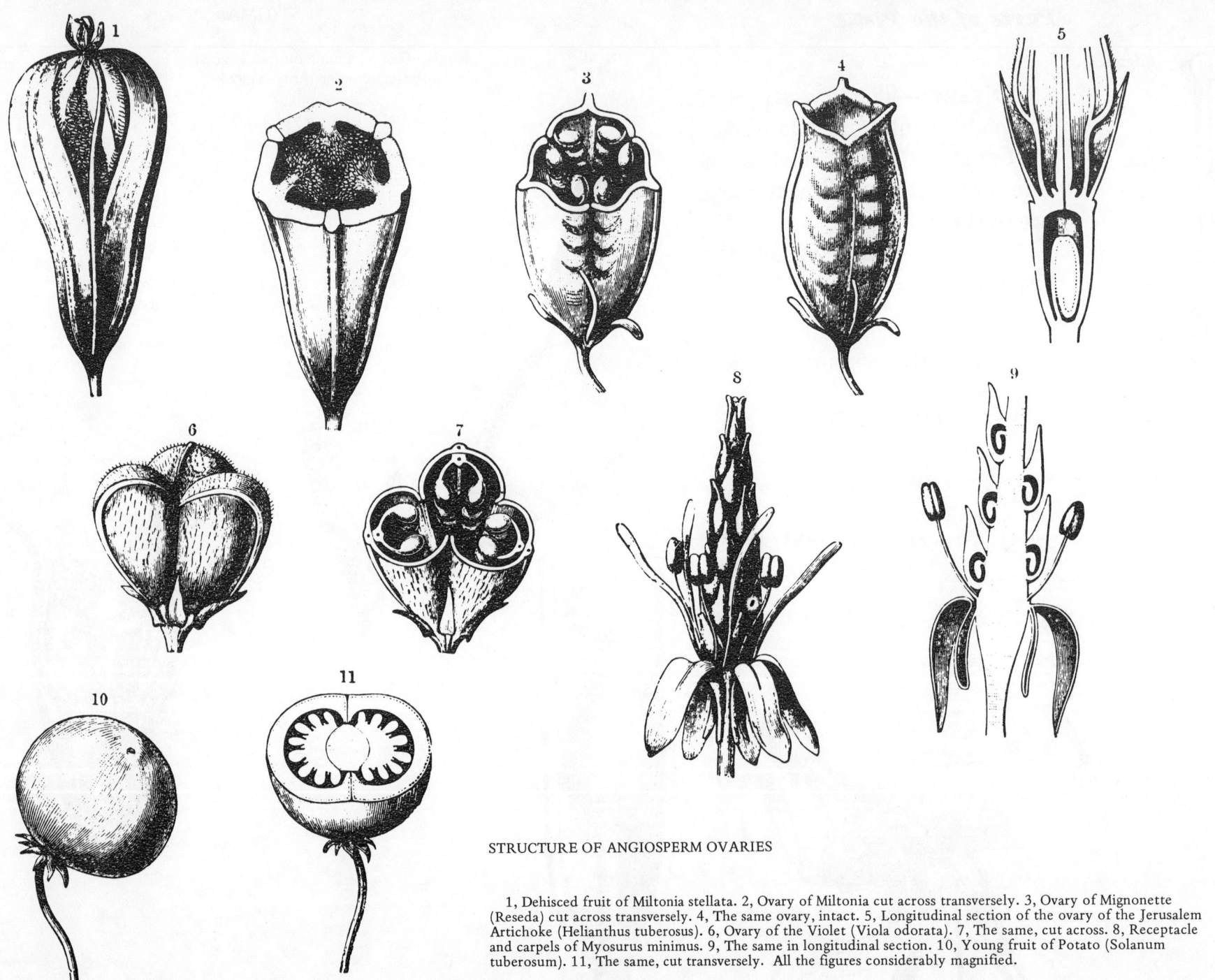

STRUCTURE OF ANGIOSPERM OVARIES

1, Dehisced fruit of Miltonia stellata. 2, Ovary of Miltonia cut across transversely. 3, Ovary of Mignonette (Reseda) cut across transversely. 4, The same ovary, intact. 5, Longitudinal section of the ovary of the Jerusalem Artichoke (Helianthus tuberosus). 6, Ovary of the Violet (Viola odorata). 7, The same, cut across. 8, Receptacle and carpels of Myosurus minimus. 9, The same in longitudinal section. 10, Young fruit of Potato (Solanum tuberosum). 11, The same, cut transversely. All the figures considerably magnified.

Parts of the Ovary

OVARY.

Parts.—Placentae, the parts to which the ovules are attached.

Dissepiments, partitions.

Cells, cavities in which the ovules are arranged.
Ovules, unfertilized seeds.

Adnation.—Inferior, calyx adherent to ovary, same as superior calyx.

Superior, calyx free from ovary, same as inferior calyx.

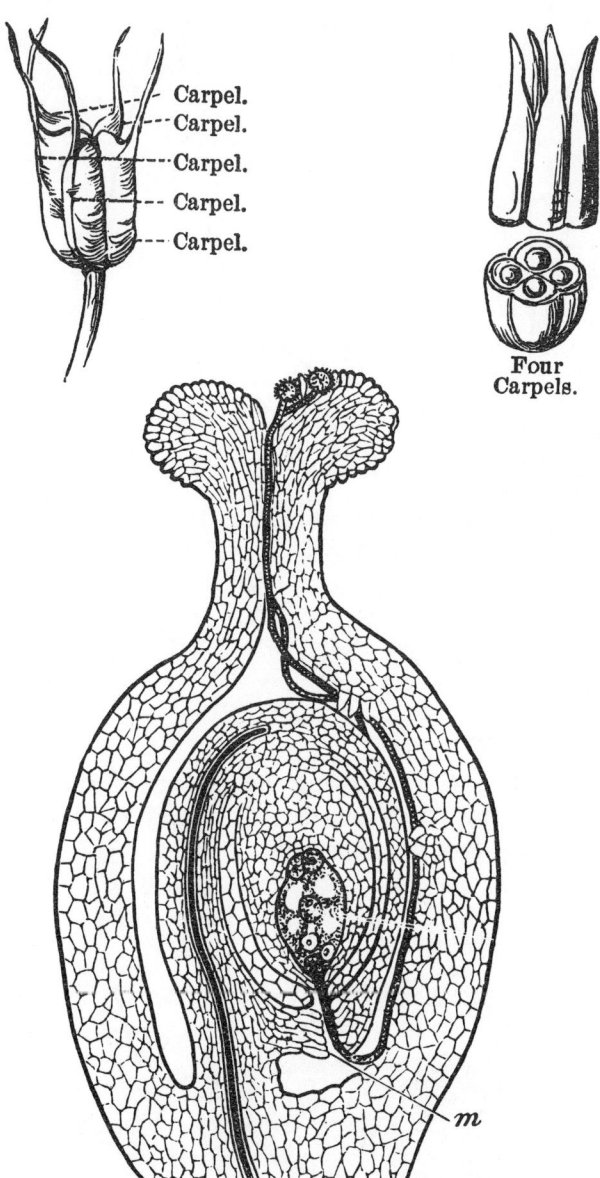

  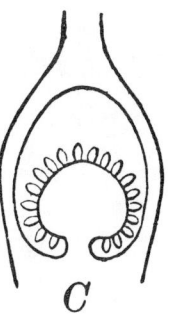

Principal types of placenta

A, parietal placenta; *B*, central placenta; *C*, free central placenta; *A* and *B*, transverse sections; *C*, longitudinal section.

Placentation.—Free-central, ovules attached to a central column in a one-celled ovary (Pink).

Axillary, ovules attached to a central column in a compound ovary.

Parietal, ovules attached to the outer walls of the ovary.

Longitudinal Section of a Pistil with a Single Ovule, showing the course of the pollen-tube to the egg-cell

m, Micropyle

OVULE.

Parts.—Nucleus, n, the essential part in which the embryo is formed.
Primine, p, the exterior coat.
Secundine, s, the interior coat.
Microphle, m, the opening of the ovary coats.
Funiculus, the stem to which the ovule is attached.
Hilum, h, the point of attachment on the ovule.
Chalaza, c, the place where the coverings and nucleus join.
Rhaphe, r, the connection between the hilum and the chalaza.
N.B.—Through the funiculus, the rhaphe, and the chalaza the ovule receives its nourishment from the placenta.

Through the micropyle it receives the tubular prolongation of the pollen.

Kinds.—Orthotropous, straight; no change in direction of parts (Buckwheat).

Campylotropous, curved; the micropyle brought near the chalaza (Bean).

Anatropous, inverted; the micropyle brought near the hilium, pointing to the placentae. Rhaphe the whole length of the ovule (Magnolia).

Amphitropous, half inverted; short rhaphe (Mallow).

Direction of Ovary.—Erect, ; ascending, ; horizontal, ; pendulous, ; suspended,

Longitudinal Section of Ovule after Fertilization

P, Pollen-tube.   n, Micropyle.   a, Outer integument.
i, Inner integument.   E, Embryo.   s, Developing endosperm.   k, Remains of tissue of ovule.   f, Stalk of ovule.

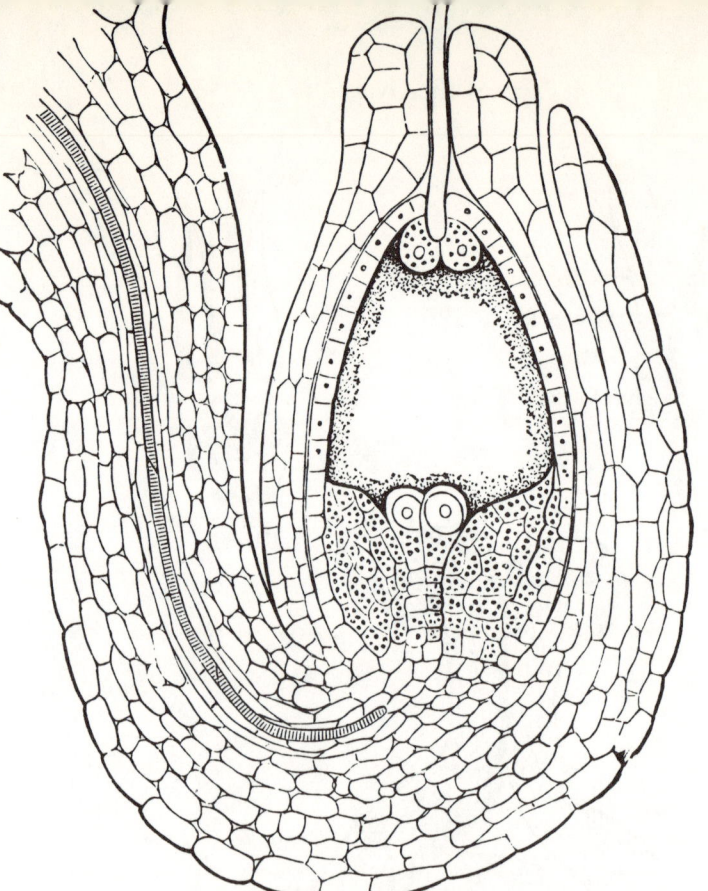

Section of an Ovule, showing the passage of the pollen-tube through the micropyle and its arrival in the neighbourhood of the egg-cell.

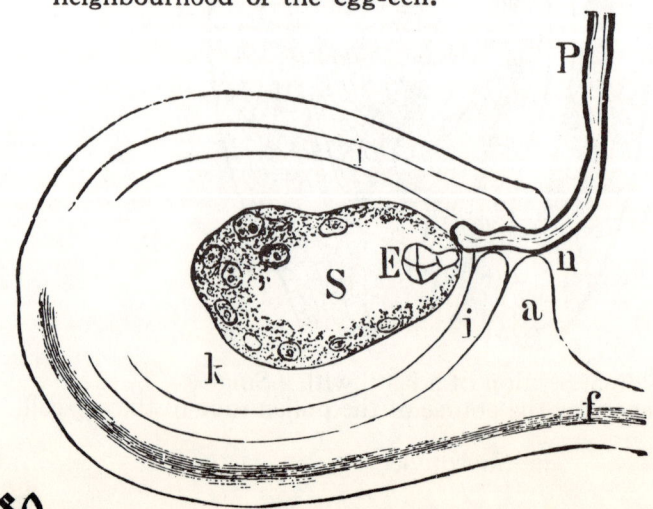

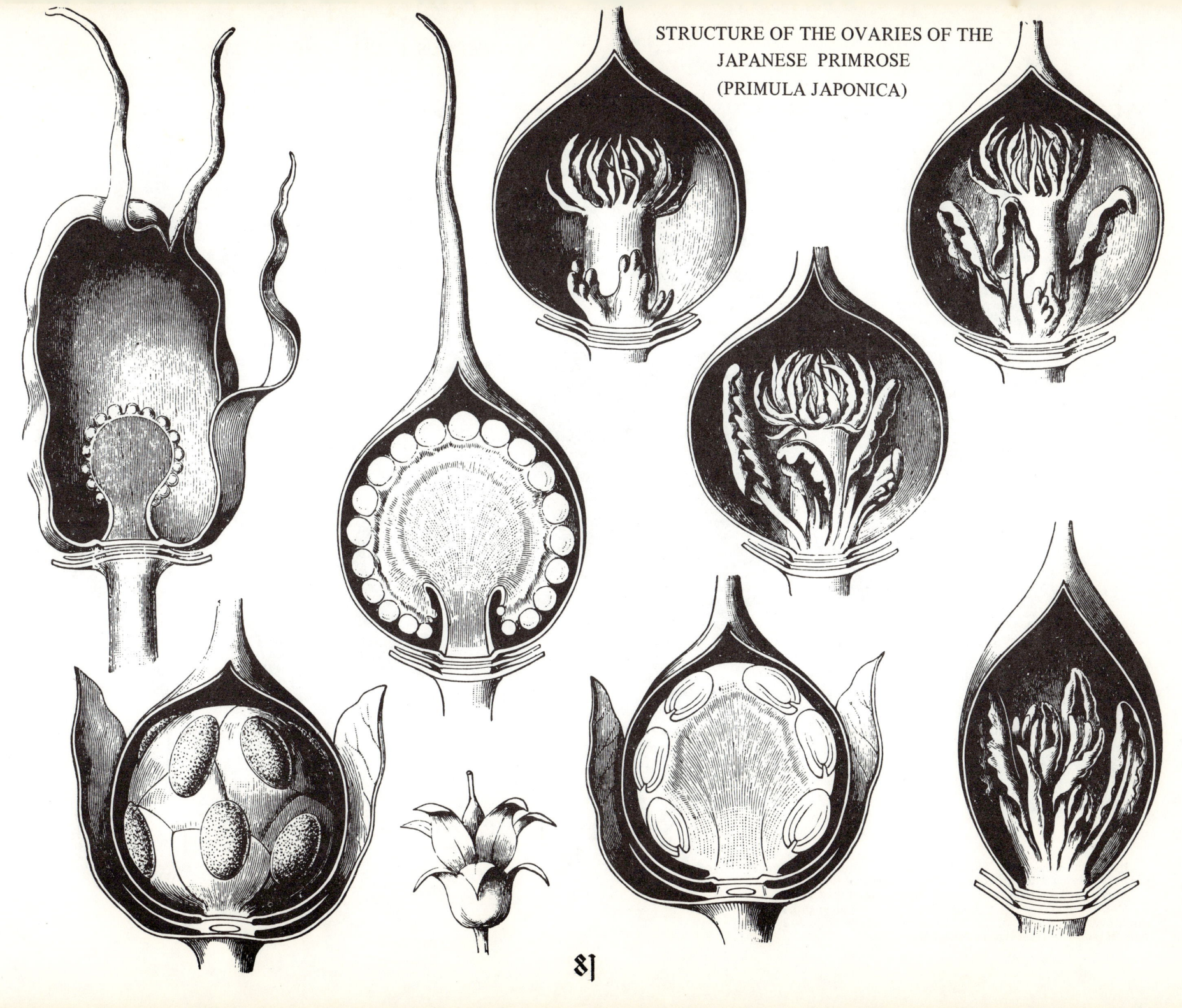

CONIFEROUS FRUITS AND SEEDS

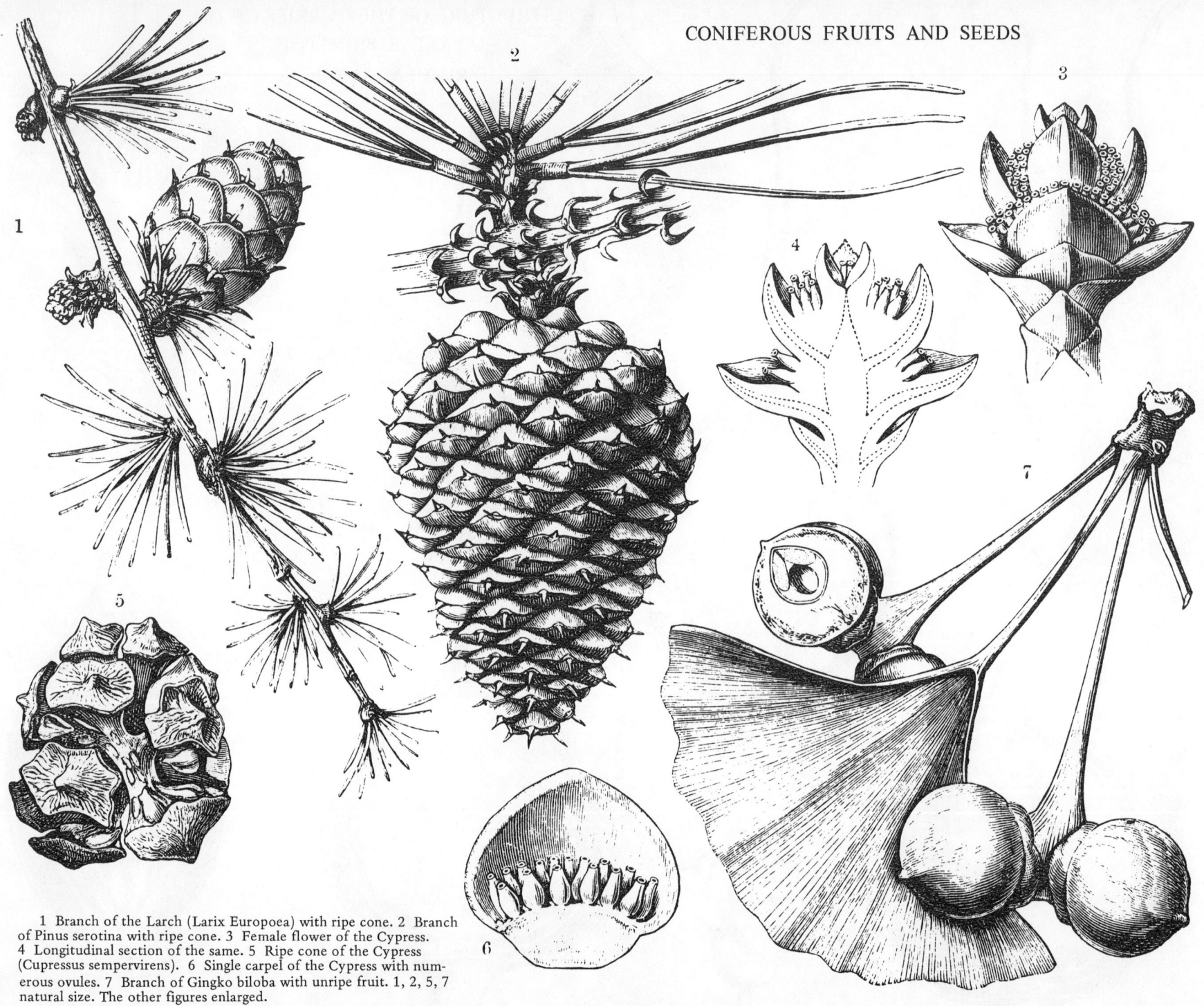

1 Branch of the Larch (Larix Europoea) with ripe cone. 2 Branch of Pinus serotina with ripe cone. 3 Female flower of the Cypress. 4 Longitudinal section of the same. 5 Ripe cone of the Cypress (Cupressus sempervirens). 6 Single carpel of the Cypress with numerous ovules. 7 Branch of Gingko biloba with unripe fruit. 1, 2, 5, 7 natural size. The other figures enlarged.

FRUIT.

Parts.—Seed, the part containing the embryo.
  Pericarp, the covering of the seeds, including the ovary and all adnate parts. The parts of the pericarp are eqicarp, or outer coat; mesocarp, or middle coat; and endocarp, or inner coat.

Dehiscence.—Septicidal, opening of the partitions.

Loculicidal, opening at the dorsal suture.

Septifragal, valves falling away from partitions.

Circumcissile, opening by a circular horizontal line.

Kinds.—Simple, aggregate, accessory, multiple.
  1. SIMPLE FRUITS.—Fleshy, Stone, Dry (formed by a single pistil).
    a. FLESHY FRUITS.—Indehiscent (with two or more seeds).

Seeds immersed in a pulpy mass.
  Berry, rind membranous (Grape).
  Hesperidium, rind leathery, separable (Orange)
  Pepo, rind hard (Cucumber).
Seeds in cells.—Pome, succulent calyx (Apple).

    b. STONE FRUITS.—Indehiscent; one-celled; endocarp hard.
  Drupe, three-coated; stone-cell entire (Peach).
  Tryma, two-coated; stone-cell two-parted (Walnut).
  Etaerio, an aggregation of drupes (Raspberry).
    c. DRY FRUITS.—Indehiscent, usually one seed with one coat.

Achenium, coat separable from seed (Dandelion).

Utricle, coat inflated (Goosefoot).
Caryopsis, coat inseparable (Wheat).

Glans, invested with a cupule, (Acorn).

Samara, having winged appendates (Maple).

c¹ DRY FRUITS.—Dehiscent.

Single pistil
  Follicle, opening by a ventral suture (Columbine).
  Legume, opening by both sutures (Bean)
  Loment, jointed legume (Desmodium).

Compound pistil
  Capsule, any compound dehiscent fruit.
  Silique, a two-valved capsule (Mustard).
  Silicle, a short silique (Shepherd's Purse).

Pyxis, circumscissile dehiscence (Purslane).

2. AGGREGATE FRUITS . A cluster of carpels on one receptacle taken as a whole (Raspberry).
3. Accessory or Anthocarpous Fruits.—Those of which the most conspicuous portion, although appearing like a pericrap in some cases, does not belong to the pistil (Rose-hip).
4. Multiple or Collective Fruits.—Those which result from the aggregation of several flowers into one mass (Pine-apple, Mulberry).
Strobile or Cone, a scaly multiple fruit, resulting from the ripening of some kinds of catkins (Hop, Conifers).
Galbalus, a closed cone (Juniper-berry, Red Cedar).

Seed.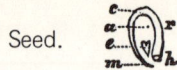

Parts.—Integuments, seed-coats. Nucleus, part containing the embryo.

1. PARTS OF INTEGUMENTS:
    Testa (episperm), the outer or proper seed-coat.
    Tegmen (endopleura), the inner coat, sometimes wanting.
    Funiculus Hilum (h), Chalaza (c), Rhaphe (r), are the same as in ovule.
    Aril, covering exterior to the integuments (not in the ovule) (May-apple, Water-lily).

    Coma, 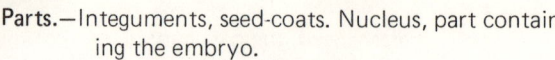 a tuft of hairs on certain seeds (Silkweed).

    This is to be distinguished from pappus, which is a tuft on the fruit (Achenium).

2. PARTS OF NUCLEUS:

    Embryo (e), the initial plantlet.

    Radicle (r), the rudimentary stem or first internode.

    Cotyledon (c), the seed leaf at the primary node.
    Plumule (p), the growing points above the cotyledons.

    Albumen (a), the food for the plantlet's first

    growth, stored outside the embryo.

GERMINATING SEEDS AND SEEDLINGS

1 Seedling of the Nasturtium (Tropoeolum majus). 2 The same at an earlier stage of development. 3 Water Chestnut (Trapo natans), from which the embryo is emerging. 4 Later stage of development. 5 Young seedling of the Austrian Oak (Quercus Austriaca). 6 The same, further developed. 7 Seed of the Date (Phoenix dactylifera) from which the embryo is emerging. 8 The same eight weeks later, after the seedling has already developed root and scale-leaves. 9 Young Date in longitudinal section. 10 Older Date in longitudinal section. 11 Seed of the Reed-mace Typha Shuttleworthii 12 The same with protruding embryo. 13 The same at a later stage of development. 14, 15 Seedling of the Sedge Carex vulgaris. Fig. 1-8, natural size; 9, 10, x 8; 11-13, x 4; 14, 15, x 6.

Kinds.—1. GENERAL FORM: Orthotropous, ; campylotropous, ; anatropous, 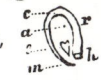; amphitropous,  same as in ovule.

2. FORM OF COVERING:
   Conformed, adhering closely to nucleus.
   Cellular, loose (Pyrola).

   Winged,  having expanded appendages (Catalpa).

   Woolly, covered closely with fibers (Cotton).

   Comose, with coma at the end (Willow Herb).

3. TEXTURE OF ALBUMEN:
   Farinaceous, mealy (Wheat).
   Oily, mealy but mixed with oil (Poppy).
   Mucilaginous, like mucilage (Morning-glory).
   Ruminated, wrinkled (Papaw).

4. NUMBER OF COTYLEDONS:

   Monocotyledonous, (Corn).

   Dicotyledonous, (Bean).

   Polycotyledonous, (Pine).

5. POSITION AND ARRANGEMENT OF EMBRYO:

   Eccentric, embryo on one side of albumen (Indian Corn).

   Peripheric, curved around albumen (Four-o'clock).

   Accumbent, applied to the cotyledons when the radicle is bent and lies along their edge (Water-cress).

   Incumbent, applied to the cotyledons when the radicle rests against the back of one of them (Shepherd's Purse).

   Conduplicate, applied to cotyledons that are incumbent and so folded as to embrace the radicle (Mustard).

6. THE DIRECTION OF THE EMBRYO AS RESPECTS THE PERICARP.
   Ascending, pointing to the apex.
   Descending, pointing to the base.
   Centripetal, pointing to the axis.
   Centrifugal, pointing to the sides.

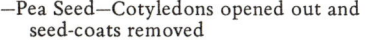

—Pea Seed—Cotyledons opened out and seed-coats removed

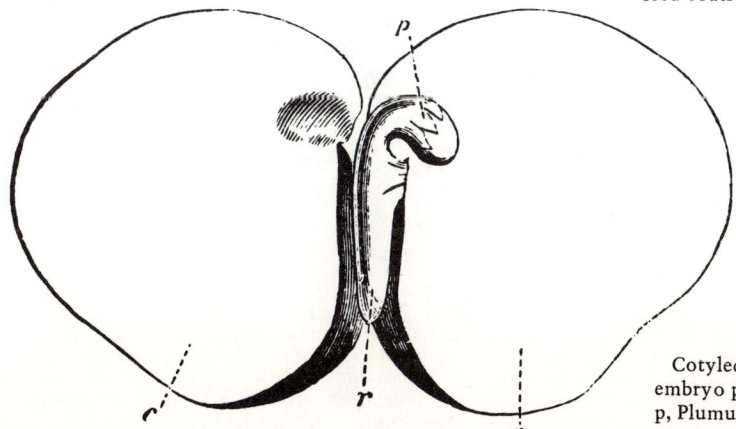

Cotyledons acting as store-places for food for the use of the embryo plant during its growth. c, c, The cotyledons. p, Plumule. r, Radicle.

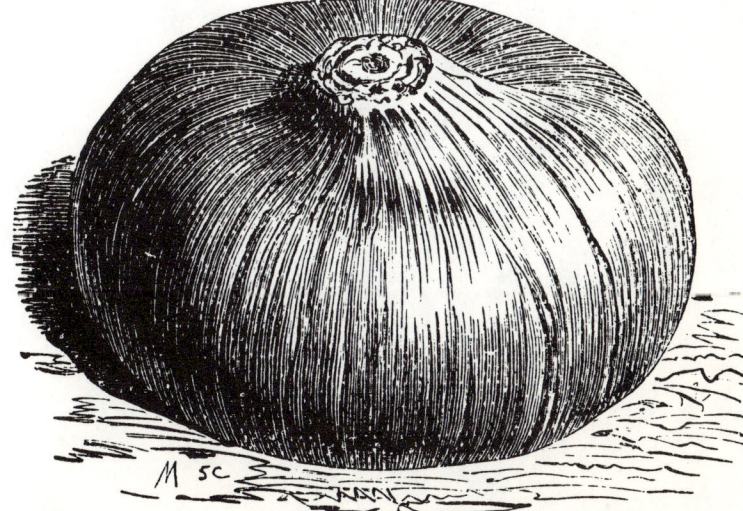

BULB OR CORM OF GLADIOLUS (UPPER PART)

POLLEN & SEED

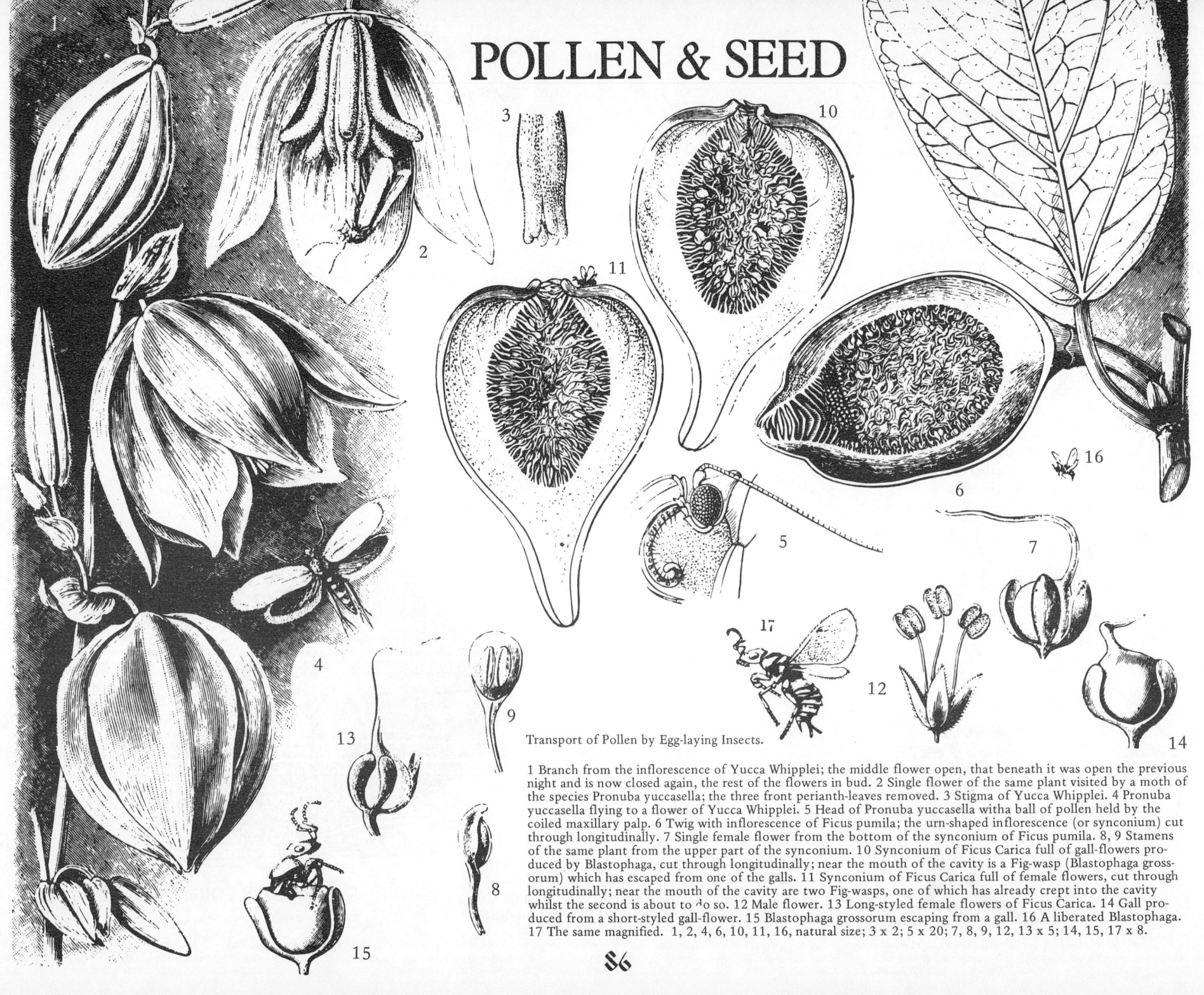

Transport of Pollen by Egg-laying Insects.

1 Branch from the inflorescence of Yucca Whipplei; the middle flower open, that beneath it was open the previous night and is now closed again, the rest of the flowers in bud. 2 Single flower of the same plant visited by a moth of the species Pronuba yuccasella; the three front perianth-leaves removed. 3 Stigma of Yucca Whipplei. 4 Pronuba yuccasella flying to a flower of Yucca Whipplei. 5 Head of Pronuba yuccasella with a ball of pollen held by the coiled maxillary palp. 6 Twig with inflorescence of Ficus pumila; the urn-shaped inflorescence (or synconium) cut through longitudinally. 7 Single female flower from the bottom of the synconium of Ficus pumila. 8, 9 Stamens of the same plant from the upper part of the synconium. 10 Synconium of Ficus Carica full of gall-flowers produced by Blastophaga, cut through longitudinally; near the mouth of the cavity is a Fig-wasp (Blastophaga grossorum) which has escaped from one of the galls. 11 Synconium of Ficus Carica full of female flowers, cut through longitudinally; near the mouth of the cavity are two Fig-wasps, one of which has already crept into the cavity whilst the second is about to do so. 12 Male flower. 13 Long-styled female flowers of Ficus Carica. 14 Gall produced from a short-styled gall-flower. 15 Blastophaga grossorum escaping from a gall. 16 A liberated Blastophaga. 17 The same magnified. 1, 2, 4, 6, 10, 11, 16, natural size; 3 x 2; 5 x 20; 7, 8, 9, 12, 13 x 5; 14, 15, 17 x 8.

PLANT SEXUAL SYSTEMS

Pollen and Seed

Seed formation involves pollination and fertilization and the production of a new, though rudimentary, plant (embryo) with stored food and a protective covering. In the angiosperms the seeds are borne within an enclosed ovary, the enlarged lower part of the pistil. When a pollen grain is carried by wind, insects, or gravity from an anther (pollen-bearing part of a stamen) to a sticky stigma (apex of a pistil) in the process of pollination, it germinates, producing a long, microscopic tube that grows down through the style into male nuclei or sperms penetrates an ovule, or rudimentary seed, and double fertilization occurs. Each ovule contains within its embryo sac eight nuclei, among them an egg nucleus and two polar nuclei. In the process of fertilization one male nucleus unites with the egg nucleus, the fertilized egg developing into the embryo, or young plant, of the seed. The other male nucleus unites with the two polar nuclei, and the fused nucleus develops into the endosperm, a food-storage tissue for the growing embryo or the young seedling which arises from it. One or two seed coats (integuments), usually thickened and hard, are formed on the outside of the ovule, but a small pore, the micropyle, remains. The mature ovule with all its parts now is a seed. The ripened ovary, containing the seeds, composed of the usually thickened ovary wall (pericarp) and any other closely associated parts, is known as a fruit. Sometimes other floral parts, such as the calyx (wintergreen) or receptacle (apple and pear), are adherent to the ovary and indistinguishable from the pericarp and are considered part of the fruit.

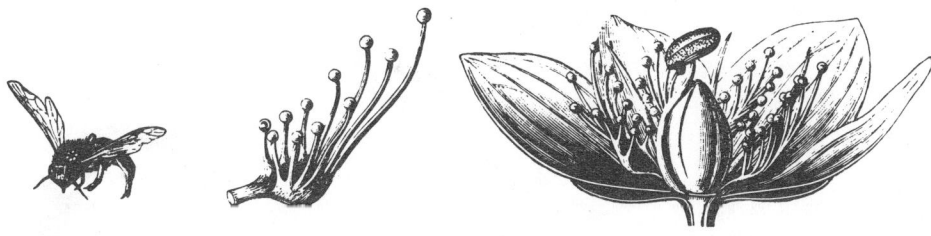

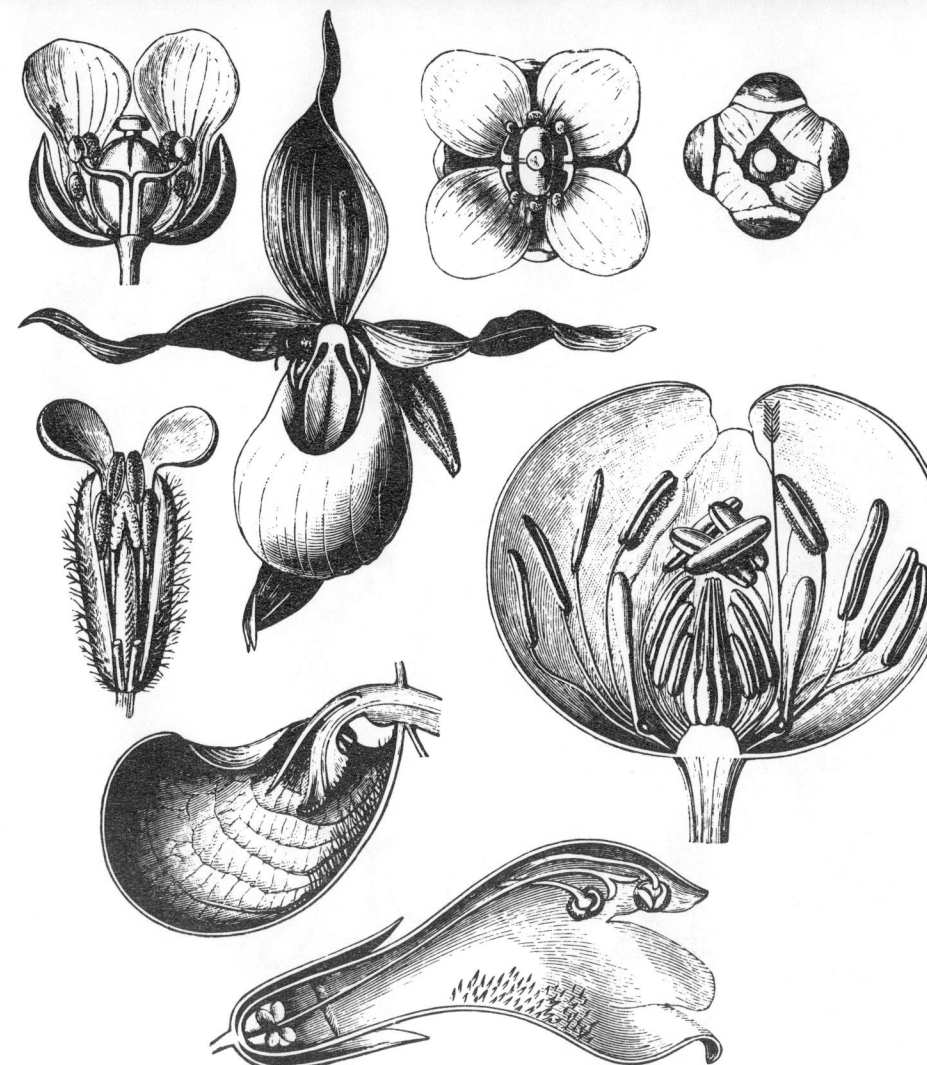

In the gymnosperms the seeds are not enclosed in a fruit but are borne loosely between the flattened scales of the ovulate cones. During pollination the wind-borne pollen grains come in direct contact with the exposed ovules. Though the minute details differ from those in angiosperms, the egg nucleus of the ovule is fertilized by a male nucleus from the pollen, the fertilized egg growing into the embryo plant. The endosperm of gymnosperms is formed directly from the ovule and not as a result of a second fertilization.

The female flower enlarges, often greatly, and becomes the familiar, hard cone (as in pines, firs, spruces), bearing the exposed or naked seeds. However, in some groups, such as junipers, the cone scales grow together to form a berrylike structure around the seeds.

In most species of trees and shrubs, fertilization occurs in the spring, and the seeds ripen relatively soon thereafter, usually in 3 to 6 months. In others, such as most of the pines, the egg is not fertilized until a year after pollination, the seed ripening in the fall of the second year. In three species, Chihuahua pine, Italian stone pine, and Torrey pine, the seeds ripen in the fall of the third year. The oaks also vary in the length of time required for seed ripening, white oaks mostly maturing their seeds in 1 year, and black oaks in 2 years.

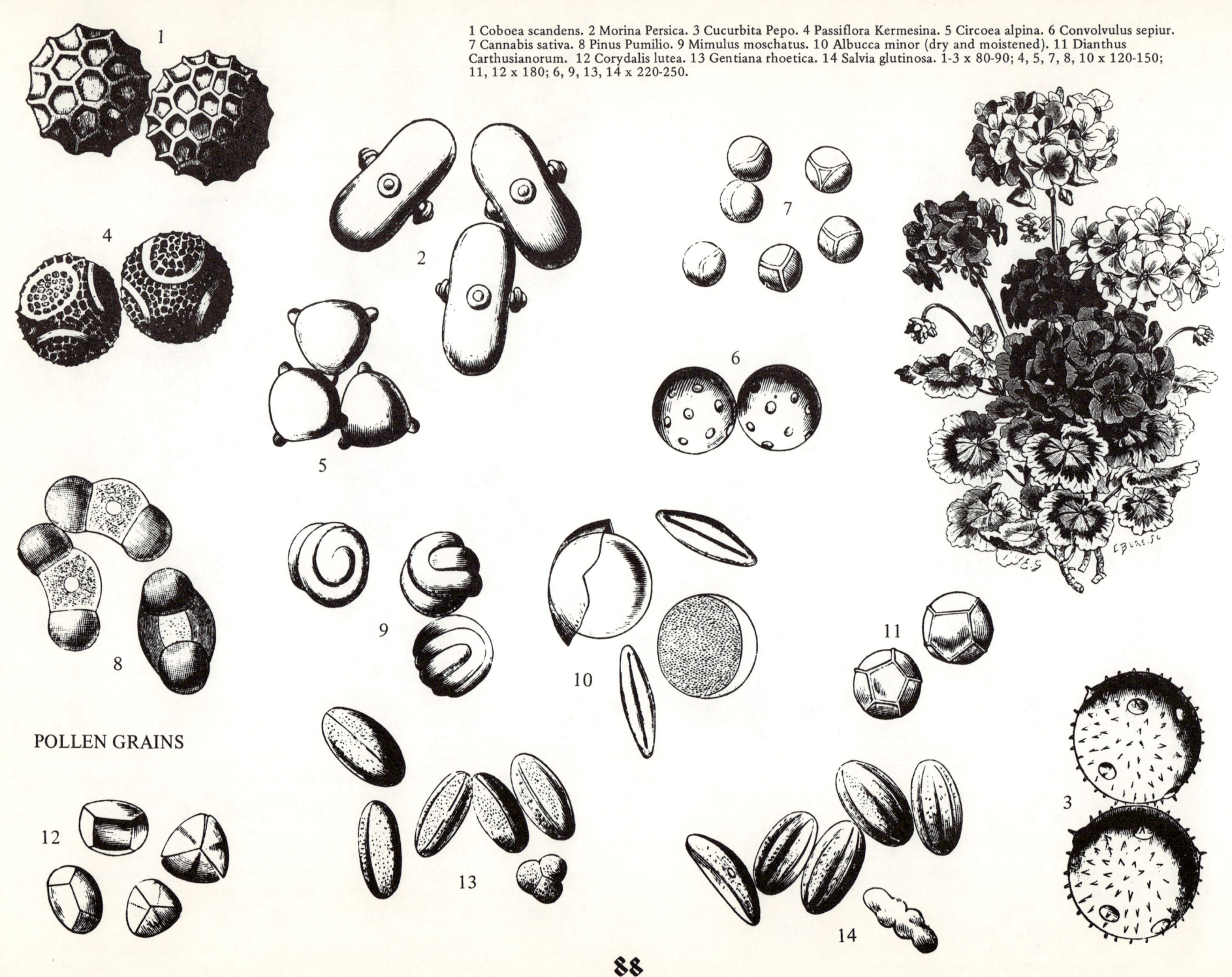

BEES, BUTTERFLIES, AND BLOSSOMS:
OUR USEFUL GARDEN INSECTS

Author Marshall D. Levin is Chief of the Apiculture Research Branch, Entomology Research Division, Agricultural Research Service.

Most of our garden plants have flowers at some time during their growing cycle. Many, of course, are ornamentals planted primarily for the flowers they produce. Many useful insects find the flowers just as attractive as we do, but for different reasons.

Flowers originally developed as showy changes to attract the attention of insects for pollination. Later modifications adapted some flowers for pollination by birds, and even bats. Many flowers returned to dependence upon the original agent of pollen dispersal, the wind. These have usually lost the eye-catching colors, forms, and odors that characterize flowers as most of us know them. Although some of these (like the grasses) are important in the garden, we grow them for other features than flowers.

Fundamental to all flowers are the floral structures required for reproduction. The female elements (stigma, style, and ovary) and the male elements (pollen, anther, and stamens) may be on the same or different plants. If on the same plant, they may be in the same or different flowers. In general, the farther apart the sexual parts are, the more dependent the plants become upon an agent of pollination to distribute the male pollen to the female pistil.

Hummingbirds and other birds provide this service for a few plants, bats also are known to pollinate some plants, but the most abundant and important pollinators are the insects that visit the flowers for food.

The pollen usually available in flowering plants provides the protein food required required by many insects, particularly the bees. It is often produced in great quantity. Many kinds of bees depend upon it for supply their young with protein, lipids, vitamins, and minerals.
Supplemented with nectar (often converted to honey), pollen thus becomes a necessity for bees, and they have evolved many remarkable structural adaptations to help them collect and handle pollen. Plants too have changed in complicated ways to take advantage of the visits of bees and other insects.

One of the most interesting of the many complicated relationships between plant and insect is that which has developed between the yucca plant of the Southwest and the yucca moth. The showy white flowers are visited by the yucca moth which purposefully scrapes pollen from the stamens and stuffs it into the funnel-shaped stigma after inserting eggs in the ovary below. This procedure guarantees food for the moth offspring which feeds on the developing ovules. The plant loses a few seeds, but is guaranteed pollination.

Other groups of insects show strong attractions to certain types of flowers. These may be roughly grouped in the following way:

Insects attracted to pollen flowers. Syrphid flies, colorful soldier flies, pollen-feeding beetles, and many pollen-collecting bees are often seen on poppy, rose, potato, elderberry, and similar flowers that provide pollen but no nectar. Male bees, moths, butterflies, or hummingbirds, interested only in nectar collection, are not usually attracted to these.

YUCCA FILAMENTOSA.

Insects attracted to flowers with exposed nectar. Short-tongued bees, flies, and many kinds of wasps are frequent visitors to the flowers of carrot, maple, saxifrage, euphorbia, poison-oak, and grapes. The flowers are usually inconspicuous, but it is easy for these insects to obtain the nectar.

Insects attracted to flowers with partly concealed nectar. Syrphid flies, short- and long-tongued bees, honey bees, and a few butterflies are attracted to the moderately showy flowers of stone fruits, strawberry, raspberry, cactus, buttercups, and cruciferous plants.

Insects attracted to flowers with concealed nectar. Many sorts of bees, wasps, and butterflies are attracted to the generally conspicuous flowers of currant, blueberry, onion, melon, and citrus. Although the nectar is hidden there is often a copious amount.

Insects attracted to social flowers. A large variety of both nectar and pollen collecting insects, including long- and short-tongued bees, showy butterflies, flies, and colorful beetles are frequent visitors to the conspicuous composites such as dandelion, sunflower, and aster. The showy "petals" are actually sterile flowers used to attract insects to the many tiny fertile florets of the central disk. The nectar in these "flowers" is usually hidden in narrow corolla tubes and the insects usually have to force their tongue past the stigma and stamens to reach the nectar.

Flowers adapted for bees. Only medium- to long-tongued bees can operate the sometimes complex mechanisms protecting the pollen and nectar of legumes, mints, sages, violets, delphinium, iris, etc. These flowers are sometimes visited by butterflies and moths for nectar, but the insects generally do not operate the pollinating mechanism. Some flowers have nectar so deep that only bumblebees can reach it. Others have tough mechanisms requiring large powerful bees for pollination. Sometimes bees will bite holes in the flowers tubes to "steal" nectar without pollinating.

Flowers adapted for butterflies and moths. Large, conspicuous, strongly perfumed flowers with nectar at the base of long narrow corolla tubes or spurs are visited principally by butterflies and moths, although some are also utilized by long-tongued bees and flies. Hummingbirds and honey birds are important pollinators in tropical areas. Examples in this group include honeysuckle, trumpet flowers, tobacco, phlox, and many orchids.

Flowers visited mostly by moths are generally open or fragrant only at night and are white or pale colored. Butterflies, on the other hand, visit flowers which are generally open and fragrant during the daytime and are variously colored.

Insect pollinators, most of which are bees, are directly beneficial to the plants they visit and as a result many are also indirectly of great benefit to man. Visits of pollinating insects to flowers usually result in the production of seeds and fruits. These seeds and fruits are very important elements in our diet and in our agricultural economy.

A partial list of crops known to require or benefit from insect pollination includes almonds, apples, cherries, cranberries, cucumbers, canteloups and watermelon, and strawberries. Lima beans, buckwheat, celery seed, mustard, rape, and sunflower are other seeds consumed by us that result from pollination.

A large number of seeds used for propagation require insect pollination. Some of the more important ones are alfalfa, asparagus, cabbage, broccoli, cauliflower, carrot, clover, onion, radish, rutabaga, and turnip. It has been conservatively estimated that bee pollination is essential to the production of $1 billion worth of agricultural crops. No one has been able to put a price tag on the value of pollination to wildflower seed production or conservation-plant maintenance.

FLOWERS OF BOMAREA CARDERI.

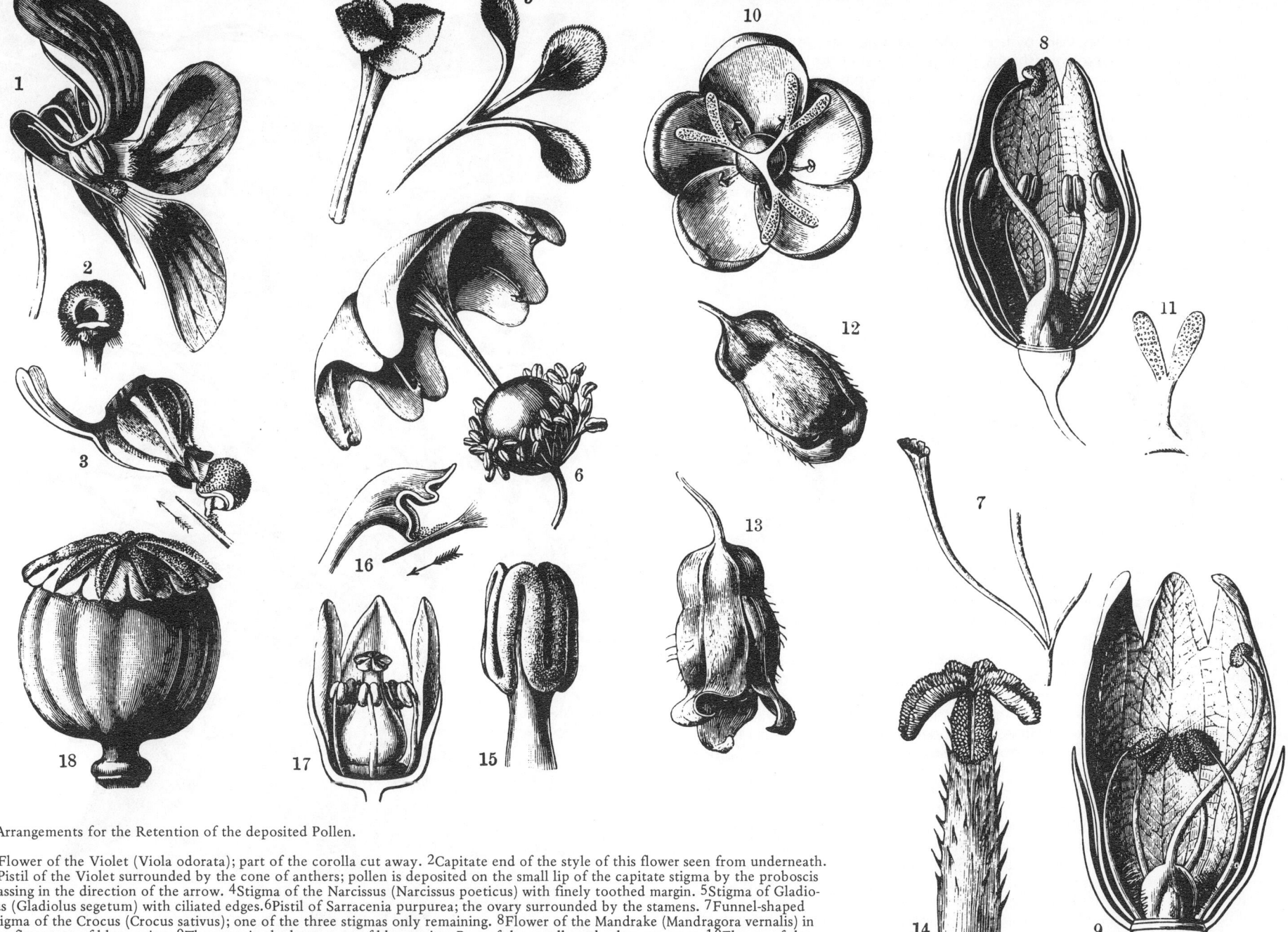

Arrangements for the Retention of the deposited Pollen.

1 Flower of the Violet (Viola odorata); part of the corolla cut away. 2 Capitate end of the style of this flower seen from underneath. 3 Pistil of the Violet surrounded by the cone of anthers; pollen is deposited on the small lip of the capitate stigma by the proboscis passing in the direction of the arrow. 4 Stigma of the Narcissus (Narcissus poeticus) with finely toothed margin. 5 Stigma of Gladiolus (Gladiolus segetum) with ciliated edges. 6 Pistil of Sarracenia purpurea; the ovary surrounded by the stamens. 7 Funnel-shaped stigma of the Crocus (Crocus sativus); one of the three stigmas only remaining. 8 Flower of the Mandrake (Mandragora vernalis) in the first stage of blossoming. 9 The same in the later stage of blossoming. Part of the corolla and calyx cut away. 10 Flower of the Sundew (Drosera longifolia) seen from above. 11 Part of the sticky papillose stigma of the Sundew. 12 Flower of the Asarabacca (Asarum Europoeum) in the first stage of blossoming. 13 The same flower at a later stage. 14 Stigma of Roemeria. 15 Stigma of Opuntia nana. 16 Stigma of Thunbergia grandiflora; pollen is deposited on the lower lip by a proboscis passing in the direction of the arrow. 17 Flower of Azalea procumbens; portions of the calyx and corolla cut away. 18 Pistil of the Opium Poppy (Papaver somniferum). 6 and 18 natural size; the others somewhat enlarged.

In your own garden, visits by honey bees and other bees to your fruit trees, holly trees, pyracantha shrubs, cucumber, muskmelon, watermelon, squash plants, blackberry, raspberry, and strawberry patches are to be greatly encouraged. They are making vital contributions to the decorative or edible fruits and berries in your garden. To prove this, put cheesecloth or other screening material around some of these plants or flower clusters so that bees cannot visit the flowers. No bees; no fruit!

Many people are disturbed at the presence of bees and wasps in their garden because of the rather common fear of being stung. It is true that bees and wasps have stings and do use them in defense of their nests. However, very rarely are their protective instincts aroused while they are visiting flowers. It is extremely unusual for anyone to be stung by a foraging bee or wasp unless the insect is sharply disturbed—accidentally or otherwise.

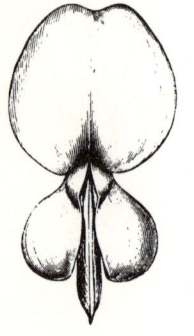 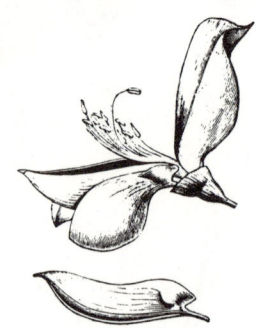 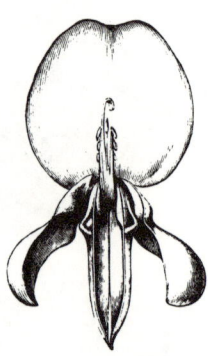

Explosive apparatus in a papillonaceous flower.

Bees or wasps around their nests are much more easily aroused to defend the nests. If you have such nests in your garden, they should either be avoided or eliminated.

One group of bees may prove annoying in another way. Leaf-cutter bees snip circular pieces of tissue from the leaves of roses and other ornamentals. They use these to fashion cells in which they store pollen and lay eggs and in which their offspring develop. Some varieties of roses are very attractive to leaf-cutter bees and are sometimes almost defoliated. However, most leaf-cutter bees are excellent pollinators and we ought to overlook the occasional damage they do to some of our plants.

Since the pollinating insects described here make such an important contribution to our food production and ecology, we should make some compromises in our attitudes towards them. Even if the ones in your garden are not helping you, they, their relatives or offspring may be making some important visits to plants in your neighbor's garden. Since bees are known to forage up to three miles, their environment encompasses a large area. Careless or uninformed use of insecticides can thus have far-reaching effects.

Some of the honey bees visiting your flowers may come from the colony of a neighbor who keeps bees as a hobby, as a 4-H or merit badge project, or for honey for his table or to sell. His bees are at the same time spreading the benefits of their activity indiscriminately around the neighborhood within a 3-mile radius. Considered in this light, we should find it easy to put into proper perspective their occasional misplaced defensive activities.

ODONTOGLOSSUM CIRRHOSUM

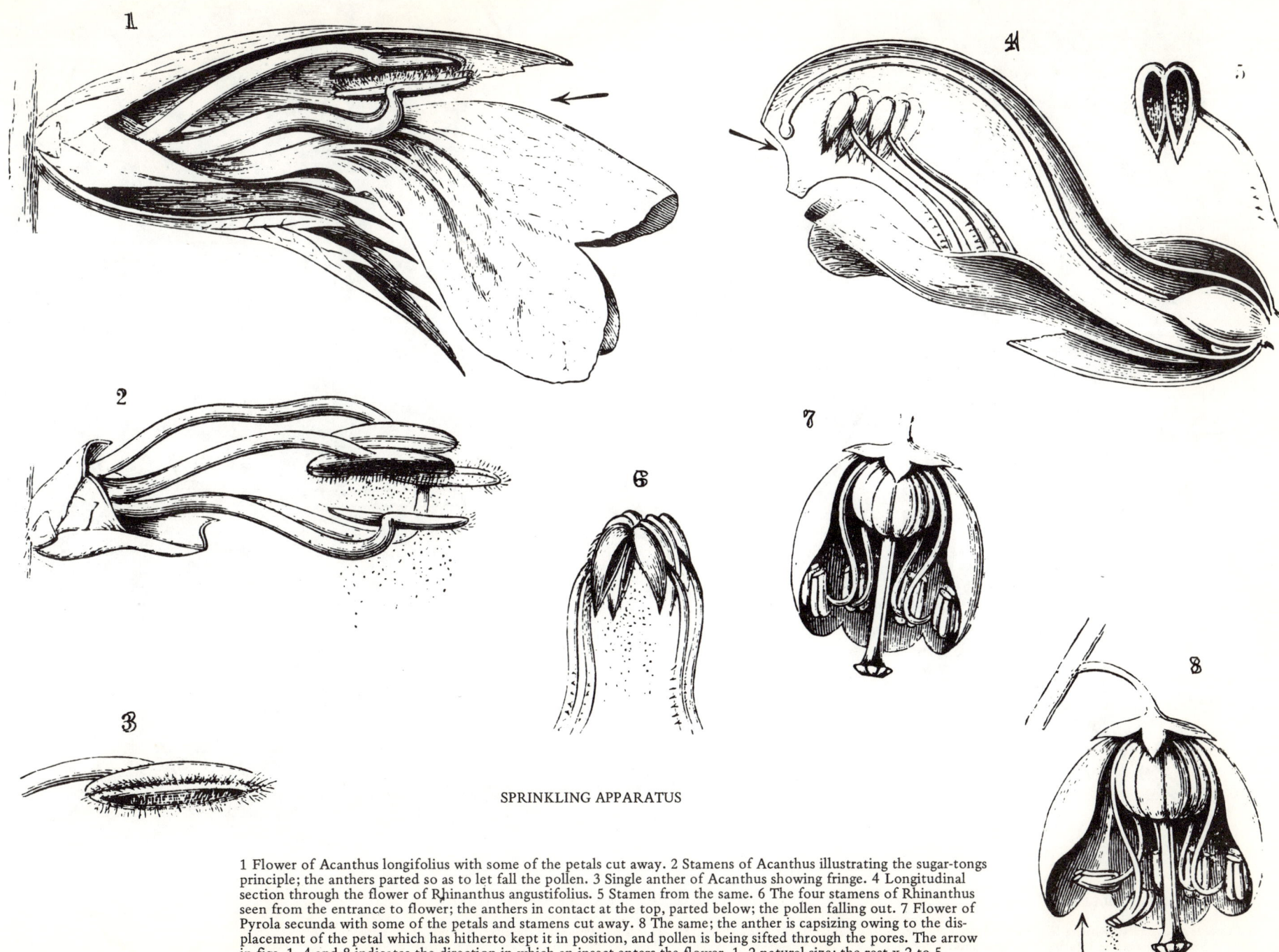

SPRINKLING APPARATUS

1 Flower of Acanthus longifolius with some of the petals cut away. 2 Stamens of Acanthus illustrating the sugar-tongs principle; the anthers parted so as to let fall the pollen. 3 Single anther of Acanthus showing fringe. 4 Longitudinal section through the flower of Rhinanthus angustifolius. 5 Stamen from the same. 6 The four stamens of Rhinanthus seen from the entrance to flower; the anthers in contact at the top, parted below; the pollen falling out. 7 Flower of Pyrola secunda with some of the petals and stamens cut away. 8 The same; the anther is capsizing owing to the displacement of the petal which has hitherto kept it in position, and pollen is being sifted through the pores. The arrow in figs. 1, 4 and 8 indicates the direction in which an insect enters the flower. 1, 2 natural size; the rest x 2 to 5.

PROTECTION OF POLLEN

1 Flowers of the Herb-Robert (Geranium Robertianum) in the daytime; the pedicels erect. 2 The same plant with its flowers pendent on curved pedicels, the position assumed during the night and in wet weather. 3 Bell-flower (Campanula patula) by day; the flower on erect pedicel. 4 Flower of the same plant inverted for the night or for wet weather, the pedicel being curved. 5 Capitulum of a Scabious (Scabiosa lucida) in the daytime; the peduncle erect. 6 Capitulum of the same plant at night or during rain, the peduncle curved and the capitulum inverted.

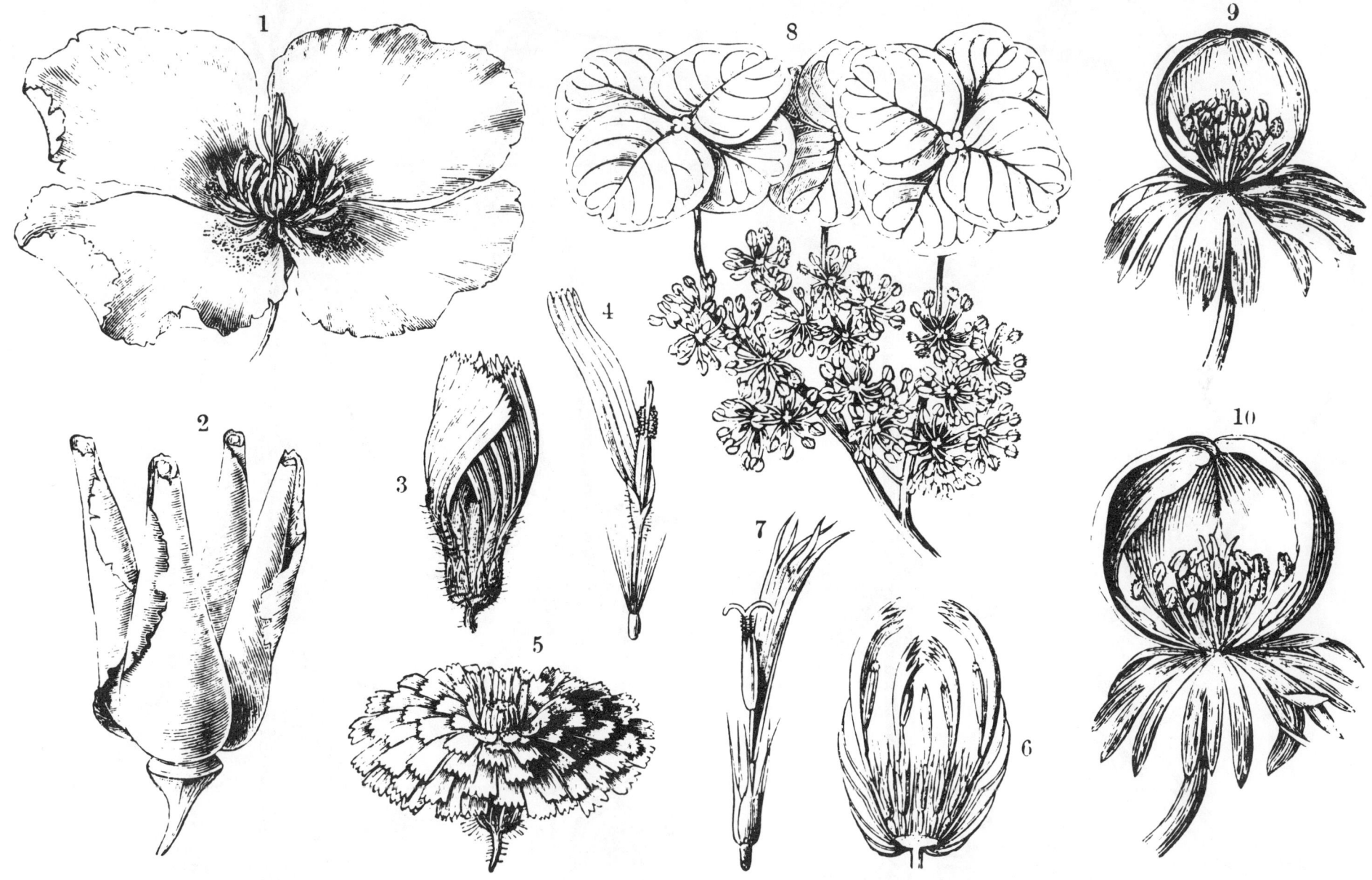

PROTECTION OF POLLEN FROM RAIN

1 Flower of Eschscholtzia Californica opened in the sunshine. 2 The same closed in wet weather. 3 Floral capitulum of Hieracium Pilosella, closed. 4 Single flower of the same plant. 5 Capitulum of the same, open. 6 Longitudinal section through a closed capitulum of Catananche coerulea. 7 Single flower taken from the capitulum in the last stage of flowering. 8 Portion of inflorescence of Hydrangea quercifolia. 9 Young closed flower of Eranthis hiemalis. 10 Old closed flower of the same.

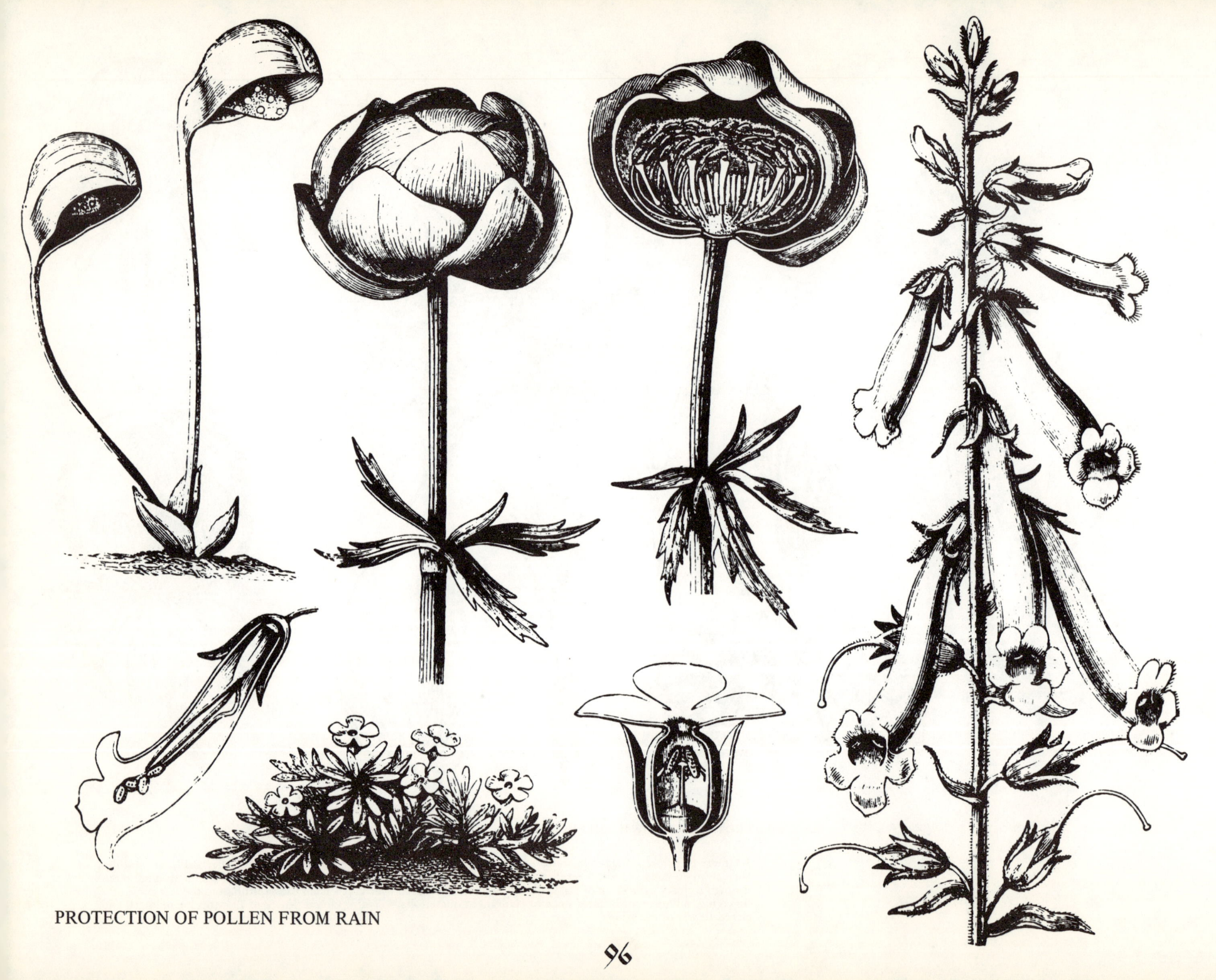

PROTECTION OF POLLEN FROM RAIN

LEFT: PROTECTION OF POLLEN FROM WET

1 Ariopsis peltata. 2 Flower of Trollius europoeus. 3 The same with some of the floral leaves removed. 4 Digitalis lutescens. 5 A single flower of Digitalis lutescens in longitudinal section. 6 Aretia glacialis. 7 Single flower of Aretia glacialis in longitudinal section (magnified).

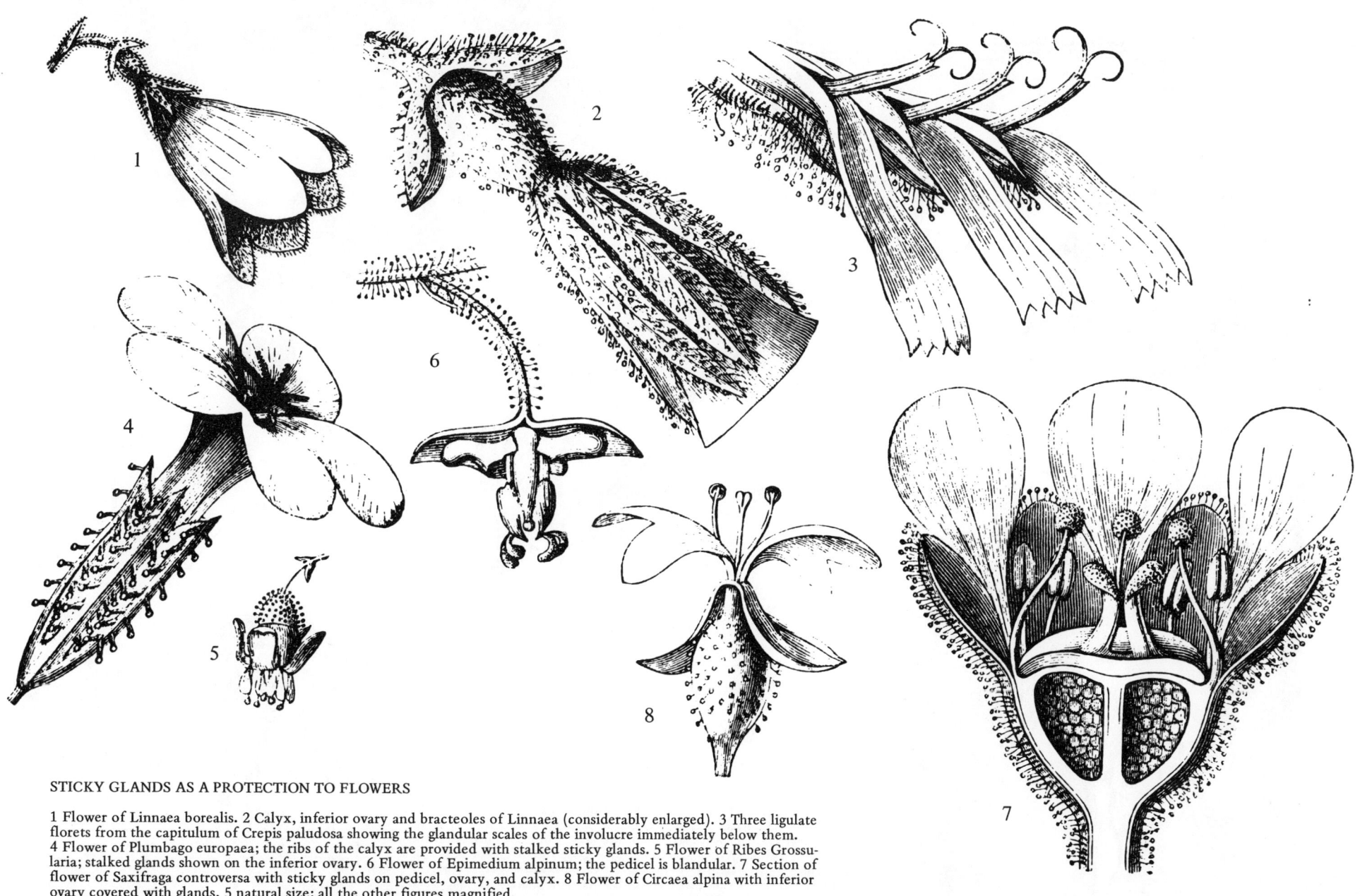

STICKY GLANDS AS A PROTECTION TO FLOWERS

1 Flower of Linnaea borealis. 2 Calyx, inferior ovary and bracteoles of Linnaea (considerably enlarged). 3 Three ligulate florets from the capitulum of Crepis paludosa showing the glandular scales of the involucre immediately below them. 4 Flower of Plumbago europaea; the ribs of the calyx are provided with stalked sticky glands. 5 Flower of Ribes Grossularia; stalked glands shown on the inferior ovary. 6 Flower of Epimedium alpinum; the pedicel is blandular. 7 Section of flower of Saxifraga controversa with sticky glands on pedicel, ovary, and calyx. 8 Flower of Circaea alpina with inferior ovary covered with glands. 5 natural size; all the other figures magnified.

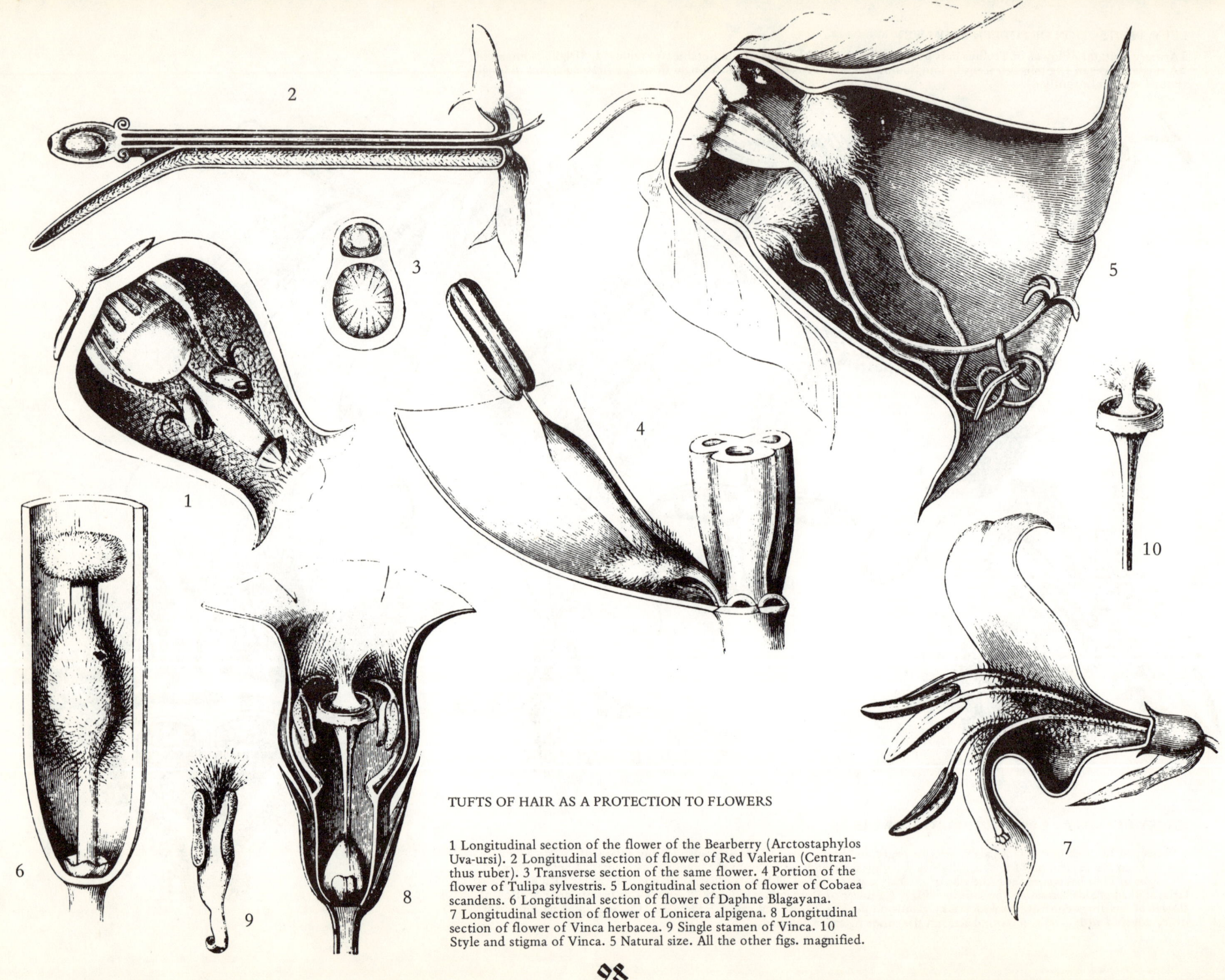

TUFTS OF HAIR AS A PROTECTION TO FLOWERS

1 Longitudinal section of the flower of the Bearberry (Arctostaphylos Uva-ursi). 2 Longitudinal section of flower of Red Valerian (Centranthus ruber). 3 Transverse section of the same flower. 4 Portion of the flower of Tulipa sylvestris. 5 Longitudinal section of flower of Cobaea scandens. 6 Longitudinal section of flower of Daphne Blagayana. 7 Longitudinal section of flower of Lonicera alpigena. 8 Longitudinal section of flower of Vinca herbacea. 9 Single stamen of Vinca. 10 Style and stigma of Vinca. 5 Natural size. All the other figs. magnified.

INTERACTIONS IN NATURE

DISPERSAL

From the standpoint of seed collection, the time of dispersal is of far greater significance than the method. The major agencies of seed dispersal are wind and animals, including birds.

Wind Dispersal

Seeds dependent on wind dispersal have a variety of structural modifications which assist flight, and form part of the seed proper (pine, spruce, fir, poplar, willows, catalpa, and the exotic Paulownia or princess tree) or of the fruit (birch, elm, ash, maple, yellow-poplar). In a few instances the instruments of flight are bracts attached to, or surrounding the fruit (basswood, hophornbeam).

Wings are the most common flight structure, associated with comparatively light seed weight. They are generally of the single-wing marginal type, as in birch and redwood seed, or the terminal type, as in ash, yellow-poplar, and pine seed. The latter type is by far the most effective in slowing down the rate of fall and thereby increasing the distance of flight. A few trees, such as some species of silverbell have four-winged fruits, which do not travel very far. In addition to wings, the hairy or cottony structures found on poplar and willow seeds are efficient for dispersal purposes. Catalpa seeds have hairy wings. A few shrubs of the family Compositae, such as Baccharis spp., produce achenes with a hairy pappus. The bladdernuts, whose fruits are inflated, buoyant capsules, exemplify a rare type of adaptation to wind dispersal among forest plants.

Some tree and shrub seeds are wind dispersed without special structural adaptations. Members of the Ericaceae, such as rhododendron and sourwood, produce extremely minute powdery seeds that are shaken out of dehiscent capsules and wafted away almost like dust. Some of the leguminous trees like redbud and locusts hold their fruits until they are torn loose by strong winds which carry the pods considerable distances. Other relatively large, light-weight fruits or fruit clusters, such as those of sweetgum and sycamore, are wind-distributed to some extent.

The distances to which fruits or seeds are carried by wind vary from a few hundred feet up to several miles. As a rule, seeds are not blown more than 100 to 500 feet from the parent stand in numbers sufficient to produce a full stand. The outstanding exceptions are willows and poplars, whose voluminous cottony seeds may be wafted in quantity over much greater distances. The seed of one species, Populus tremula, has been observed to travel 1,700 feet. Slope may markedly affect the distance of dispersal. Theoretically, a seed floating down to a 45 degree angle to level ground will travel a distance equal to the height of the tree. On a 30 degree slope, however, a seed falling at a 45 degree angle will travel 2.69 times as far down slope as on level ground before coming to earth, but only 0.73 as far up slope.

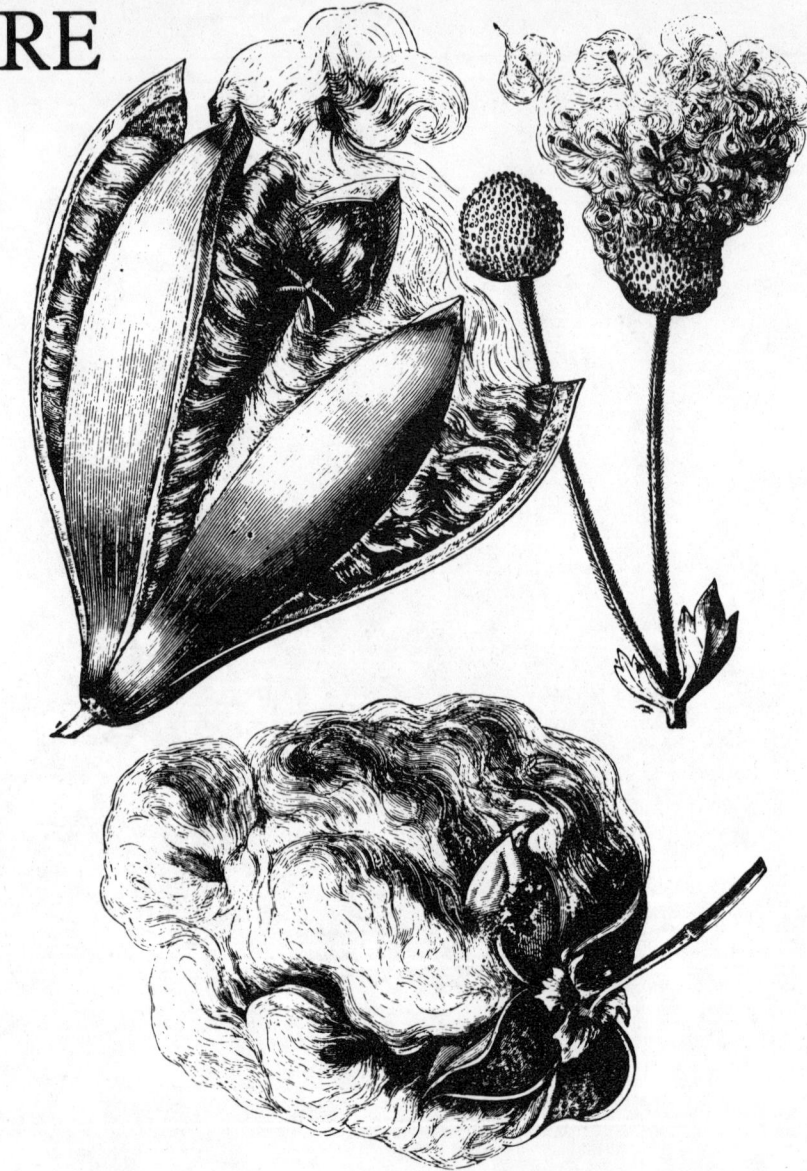

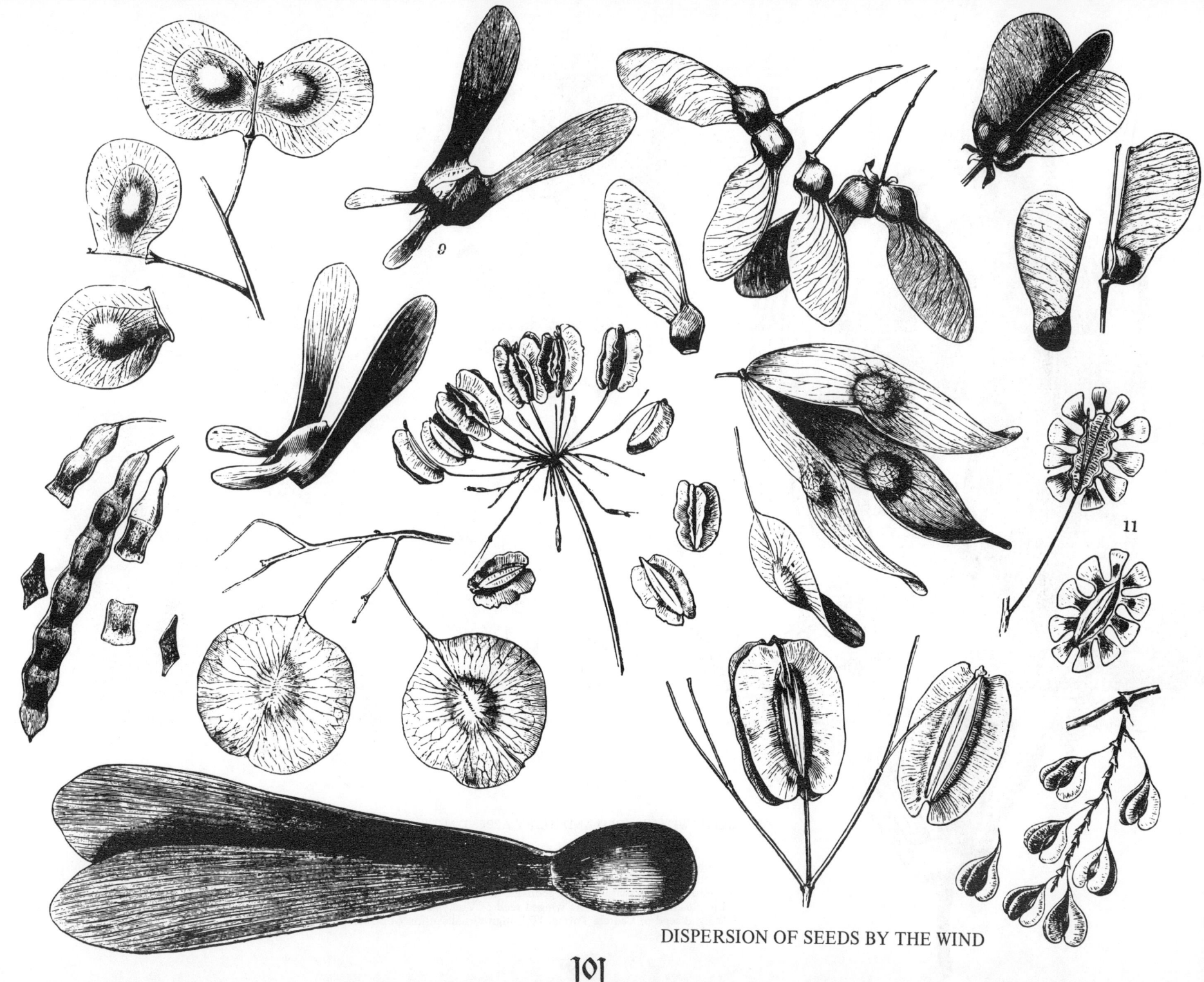

DISPERSION OF SEEDS BY THE WIND

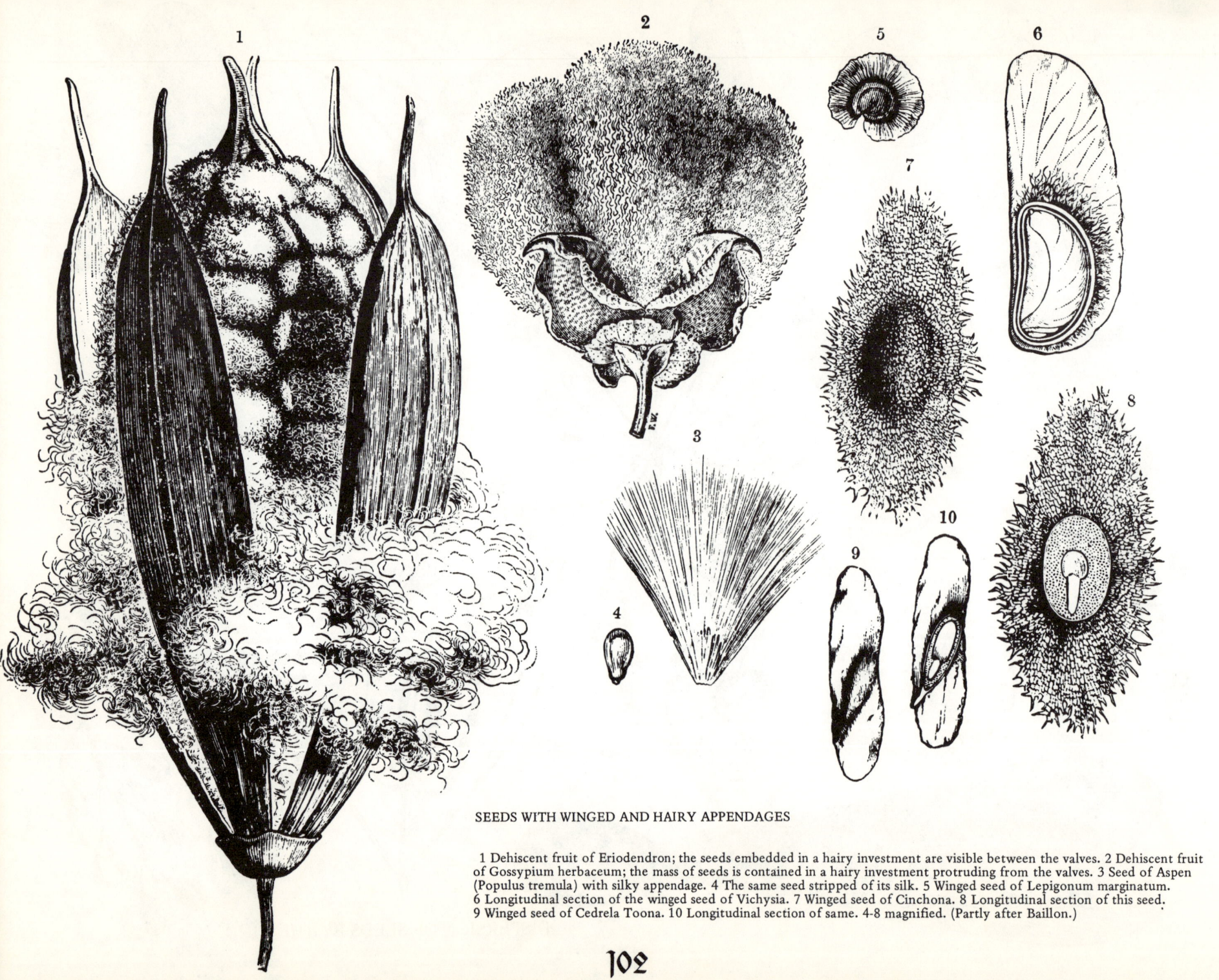

SEEDS WITH WINGED AND HAIRY APPENDAGES

1 Dehiscent fruit of Eriodendron; the seeds embedded in a hairy investment are visible between the valves. 2 Dehiscent fruit of Gossypium herbaceum; the mass of seeds is contained in a hairy investment protruding from the valves. 3 Seed of Aspen (Populus tremula) with silky appendage. 4 The same seed stripped of its silk. 5 Winged seed of Lepigonum marginatum. 6 Longitudinal section of the winged seed of Vichysia. 7 Winged seed of Cinchona. 8 Longitudinal section of this seed. 9 Winged seed of Cedrela Toona. 10 Longitudinal section of same. 4-8 magnified. (Partly after Baillon.)

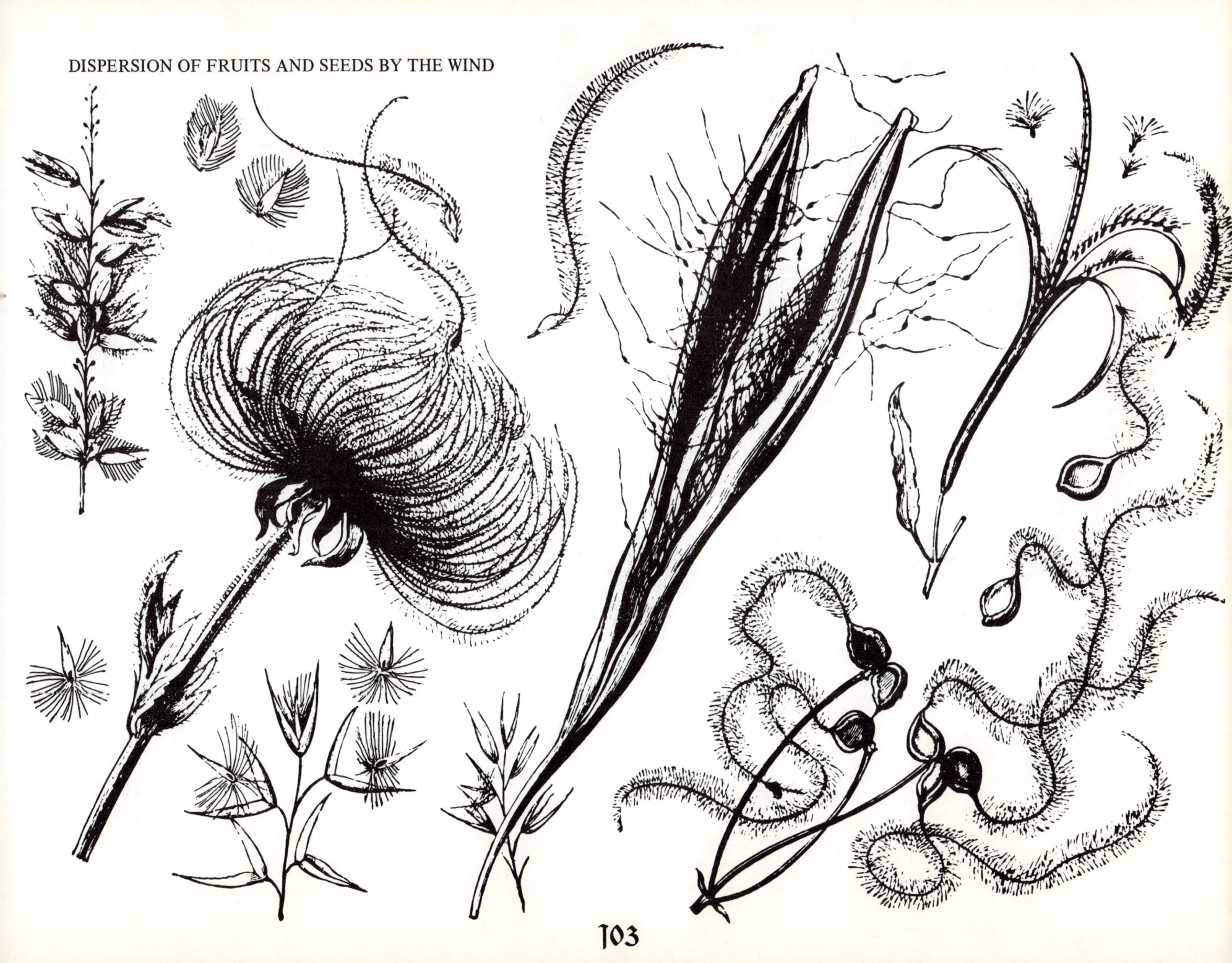

DISPERSION OF FRUITS AND SEEDS BY THE WIND

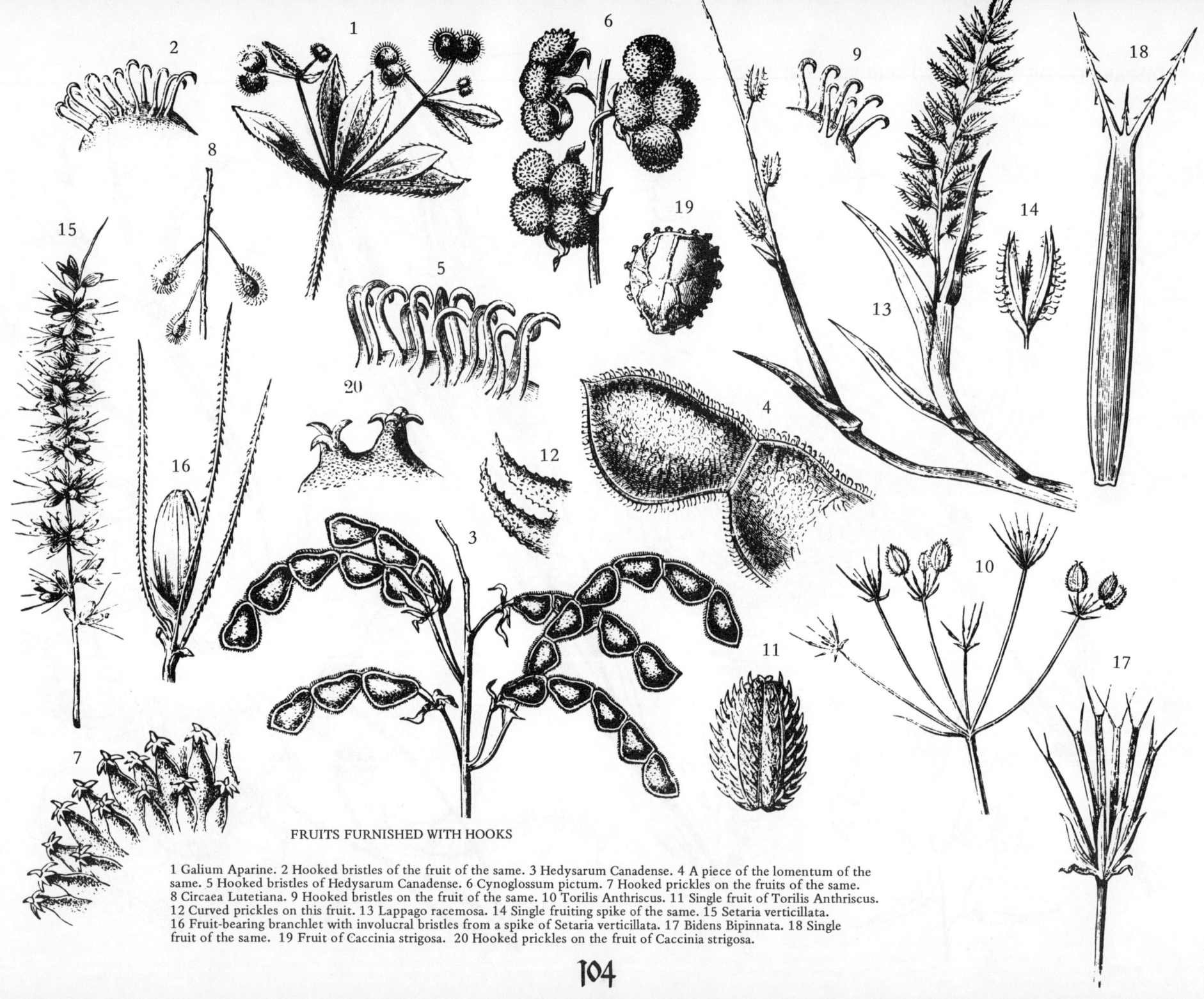

FRUITS FURNISHED WITH HOOKS

1 Galium Aparine. 2 Hooked bristles of the fruit of the same. 3 Hedysarum Canadense. 4 A piece of the lomentum of the same. 5 Hooked bristles of Hedysarum Canadense. 6 Cynoglossum pictum. 7 Hooked prickles on the fruits of the same. 8 Circaea Lutetiana. 9 Hooked bristles on the fruit of the same. 10 Torilis Anthriscus. 11 Single fruit of Torilis Anthriscus. 12 Curved prickles on this fruit. 13 Lappago racemosa. 14 Single fruiting spike of the same. 15 Setaria verticillata. 16 Fruit-bearing branchlet with involucral bristles from a spike of Setaria verticillata. 17 Bidens Bipinnata. 18 Single fruit of the same. 19 Fruit of Caccinia strigosa. 20 Hooked prickles on the fruit of Caccinia strigosa.

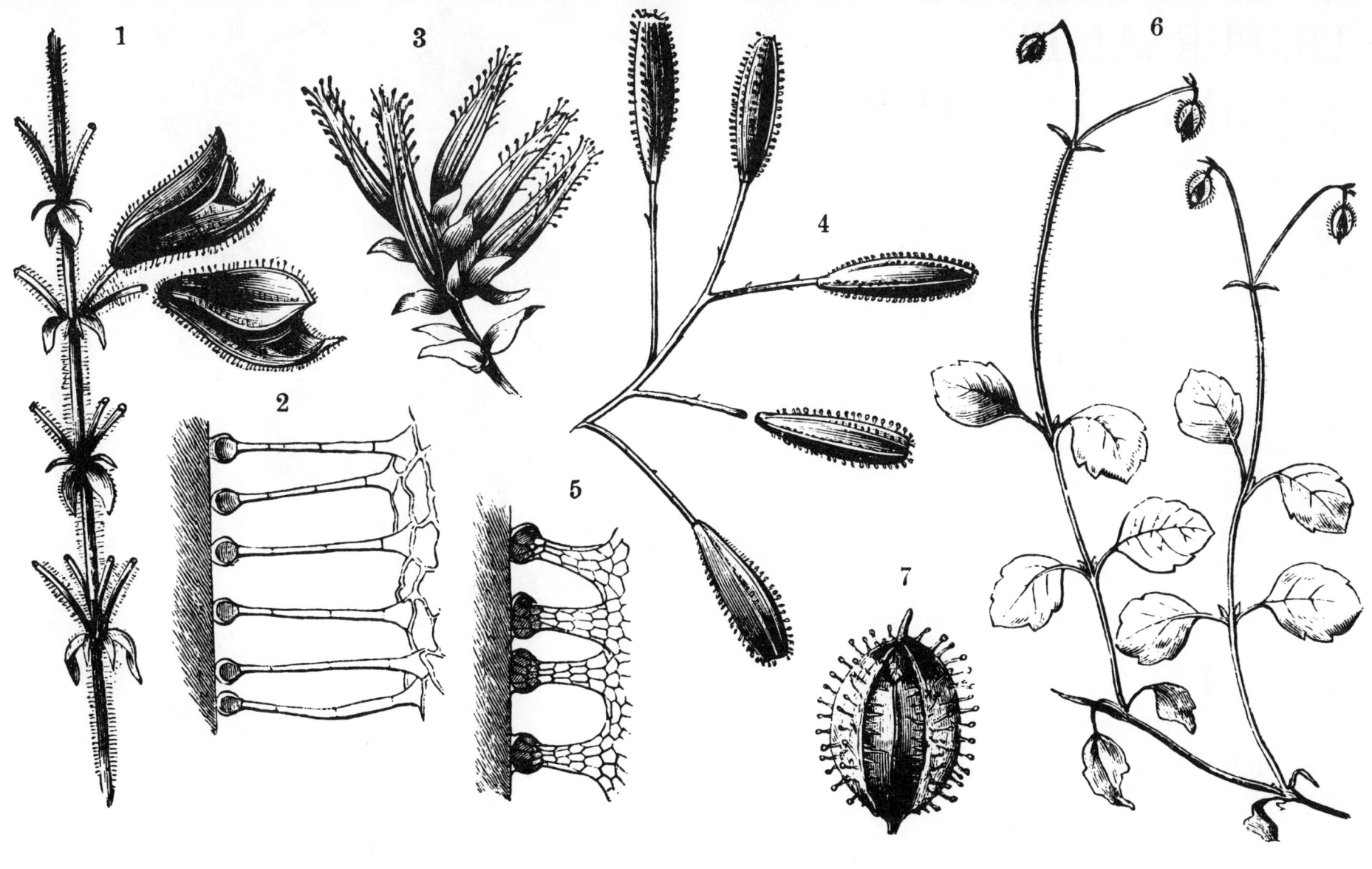

STICKY FRUITS

1 Salvia glutinosa. 2 Stalked adhesive glands on the fruiting calyx of the same; 3 Plumbago Capensis. 4 Pisonia aculeata. 5 Stalked adhesive glands on the fruit of the same. 6 Linnaea borealis. 7 Fruit of the same.

DISPERSAL BY MAMMALS & BIRDS

The seeds of forest trees and shrubs do not display so many structural adaptations as are found among seeds of herbaceous species for distribution by animals. Bur fruits and barbed or hooked appendages are rare, and even the burs on the seeds of such species as chestnut and chinquapin are not of much aid in distribution. Sticky seeds or fruits which adhere to mammals or birds are also uncommon, except those of the parasitic mistletoes, apparently carried by birds.

The most important means of distribution by animals in the forest are (1) the eating of fleshy fruits with hard seed which pass intact through their digestive tracts, and (2) hoarding. Birds eat and distribute most of the small-seeded berries and berrylike fruits though certain mammals such as bears and skunks also help.

Aldous reports that Vaccinium seed have been germinated from bear dung and rose seed from grouse droppings, and that seeds of Chiogenes, Rubus, and Vaccinium passed through chipmunks apparently intact, though unable to germinate. Tests showed that seed of the following species gave reasonably good germination after passage through birds: Blackcap raspberry, blackberry, Missouri gooseberry, American elder, Tatarian honeysuckle, black cherry, poison sumac, and meadow rose. There is some evidence that passage of juniper seed through birds is an aid to germination. Presumably, the grinding action in the bird's crop, or enzyme action in the digestive tract, renders the seed coats permeable. Birds often carry fruits such as cherries and plums to a convenient perch, consume them and discard the seed.

Dispersal of seed by hoarding is effected primarily by squirrels, and to some extent by mice and birds. Squirrels hoard mostly cones and nut fruits; mice collect conifer and other comparatively small seed. Usually some of the hoarded seed is untouched during the winter and germinates the following spring. Nuts require moist conditions for survival during winter, and burial by squirrels under litter is an excellent storage and planting method.

The hoarding habits of animals are vital to the forest succession of nut trees. Most of these species are somewhat tolerant; they tend to follow and replace pioneer trees, such as certain pines, aspens, and pin cherry that typically spring up after fire or on old fields. However, the invasion of pioneer forest communities by oaks, hickories, and similar species depends almost entirely on the hoarding activities of animals—primarily squirrels. Without this hoarding plant succession in many forest communities would be slowed down, and might actually be different.

The Common Squirrel.

PROTECTION OF RIPENING SEEDS AGAINST THE ATTACK OF ANIMALS

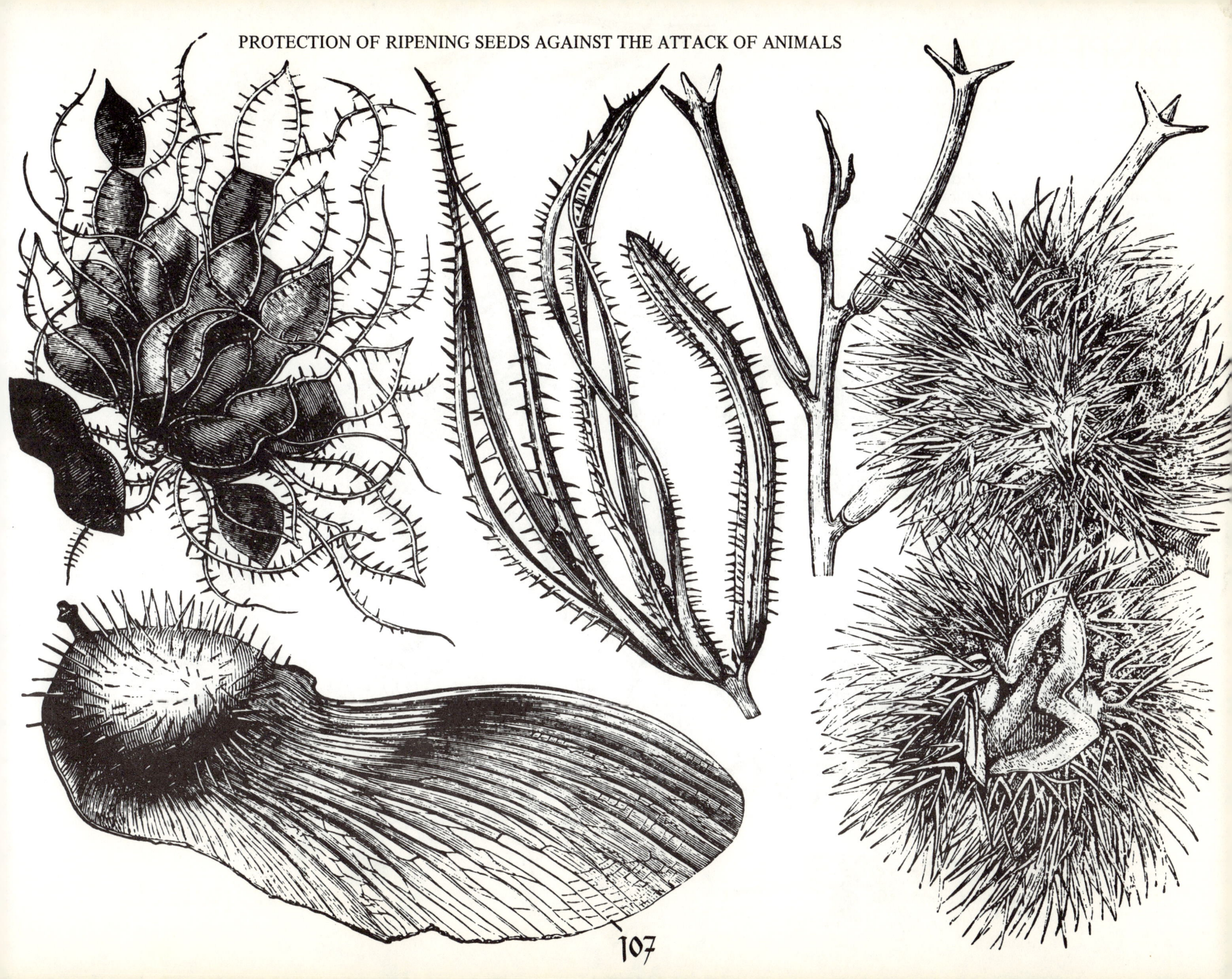

107

Other Methods of Dispersal

Gravity is important in seed dispersal on steep slopes, especially with heavy, globose fruits such as acorns, walnuts, apples, and persimmons. Running water may be a factor occasionally, but is important in the forest only for flood-plain species like willows, alders, red birch, sycamore, etc., and the seeds of even these species are adapted to other methods of dispersal. Landslides or snowslides, rain and ocean currents occasionally help to distribute seeds. The forceful ejection of seeds from the fruit is rare in trees, most common in herbaceous species, and occasionally found in shrubs, of which the witchhazel is a classic example. The Para rubber tree (Hevea brasiliensis) is one of the few forest species bearing explosive fruits.

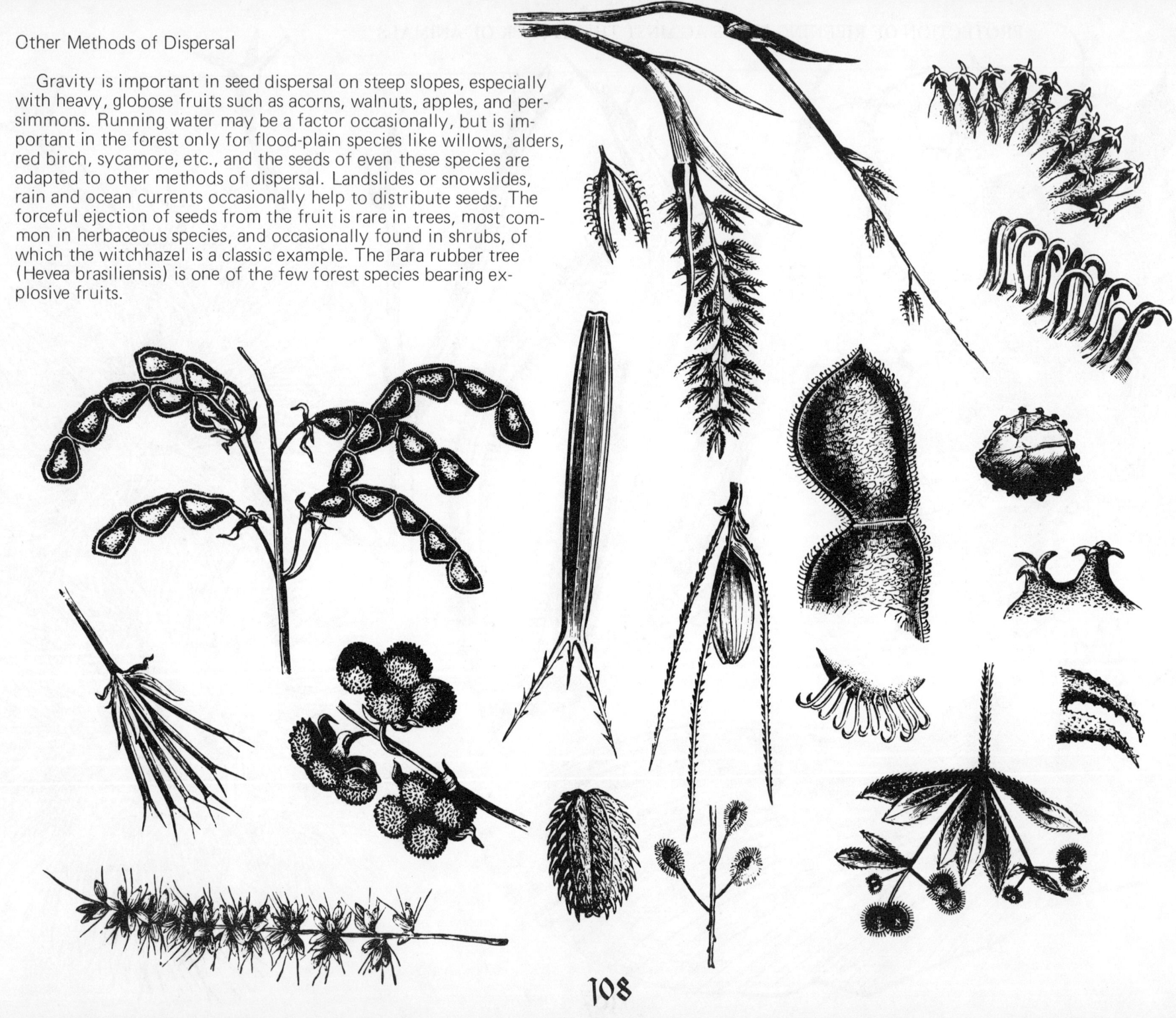

1 Umbellate raceme of Lobularia nummularioefolia with flowers and young fruits. 2 A single young flower of the same plant. 3 A young fruit of the same plant with two of the enlarged white petals attached to it. 4 Flower spike of Lavandula Stoechas ending in a crest of empty blue bracts. 5 Umbellate raceme of Alyssum cuneatum with young flat open flowers in the center and old closed flowers at the circumference. 6 Petal of a young flatly-opened flower of the same plant. 7 Petal of an old closed flower of the same plant. 8 Raceme of Muscari comosum; the upper long-stalked flowers crowded into a head are sterile. 9 Inflorescence of Trifolium badium; the upper young flowers are light yellow, the old lower drooping flowers are dark brown. 10 A branch from the inflorescence of Halimocnemis mollissima; the erect bladder-like appendages of the anthers protrude from the insignificant perianth and look like petals. 11 A single stamen of Halimocnemis mollissima; the connective rises above the anther in the form of a bladder-shaped appendage. 12 Inflorescence of Cornus florida surrounded by four large white bracts. 13 Cornflower (Centaurea Cyanus); the small flowers of the disc are surrounded by large funnel-shaped sterile flowers. 14 Raceme of Kernera saxatilis; the ovaries in the center of the old flowers are darkly colored and surrounded by the enlarged petals. 15 Inflorescence of the umbelliferous Orlaya grandiflora; the peripheral flowers radiate outwards. 16 A single radiating flower of the same plant. 17 Umbellate raceme of the Candytuft (Iberis amara); the outwardly-directed petals of the peripheral flowers are twice as large as those which are turned towards the center of the inflorescence.

Concealment of Honey

There are two kinds of contrivances for hiding the honey in pits, tubes, and channels. In the one the entrance to the hiding-place is narrowed by all kinds of inflations, cushions, bands, and flaps at the mouth of the flower-tube (see above figure). In the other the nectary is completely closed over by a roof or door, or by two lips, so that those animals which desire the honey stowed away in the cavity are compelled either to raise the roof, to open the door, or to press down one of the lips.

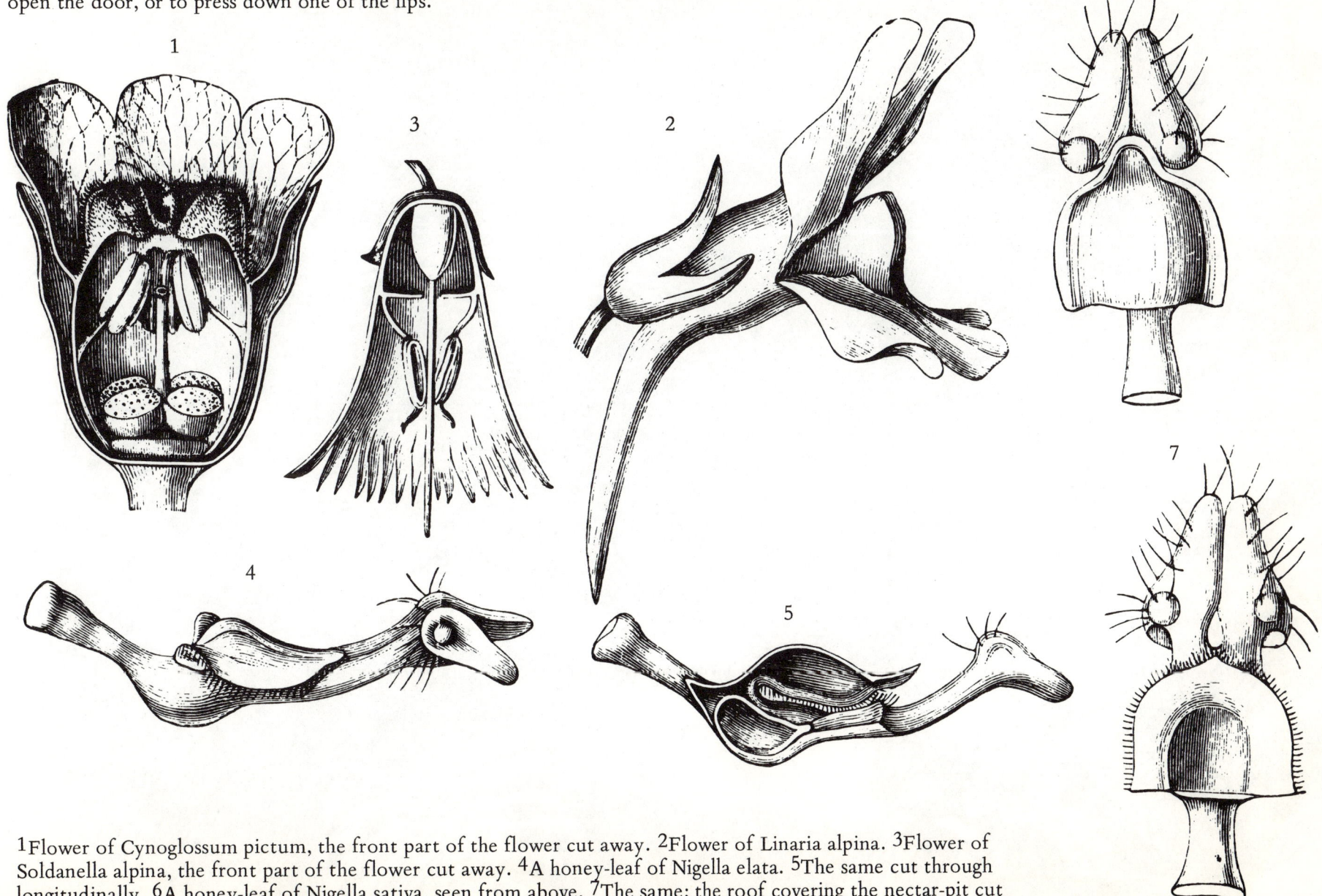

¹Flower of Cynoglossum pictum, the front part of the flower cut away. ²Flower of Linaria alpina. ³Flower of Soldanella alpina, the front part of the flower cut away. ⁴A honey-leaf of Nigella elata. ⁵The same cut through longitudinally. ⁶A honey-leaf of Nigella sativa, seen from above. ⁷The same; the roof covering the nectar-pit cut away. All the figures somewhat enlarged.

FANTASIES OF FERNS

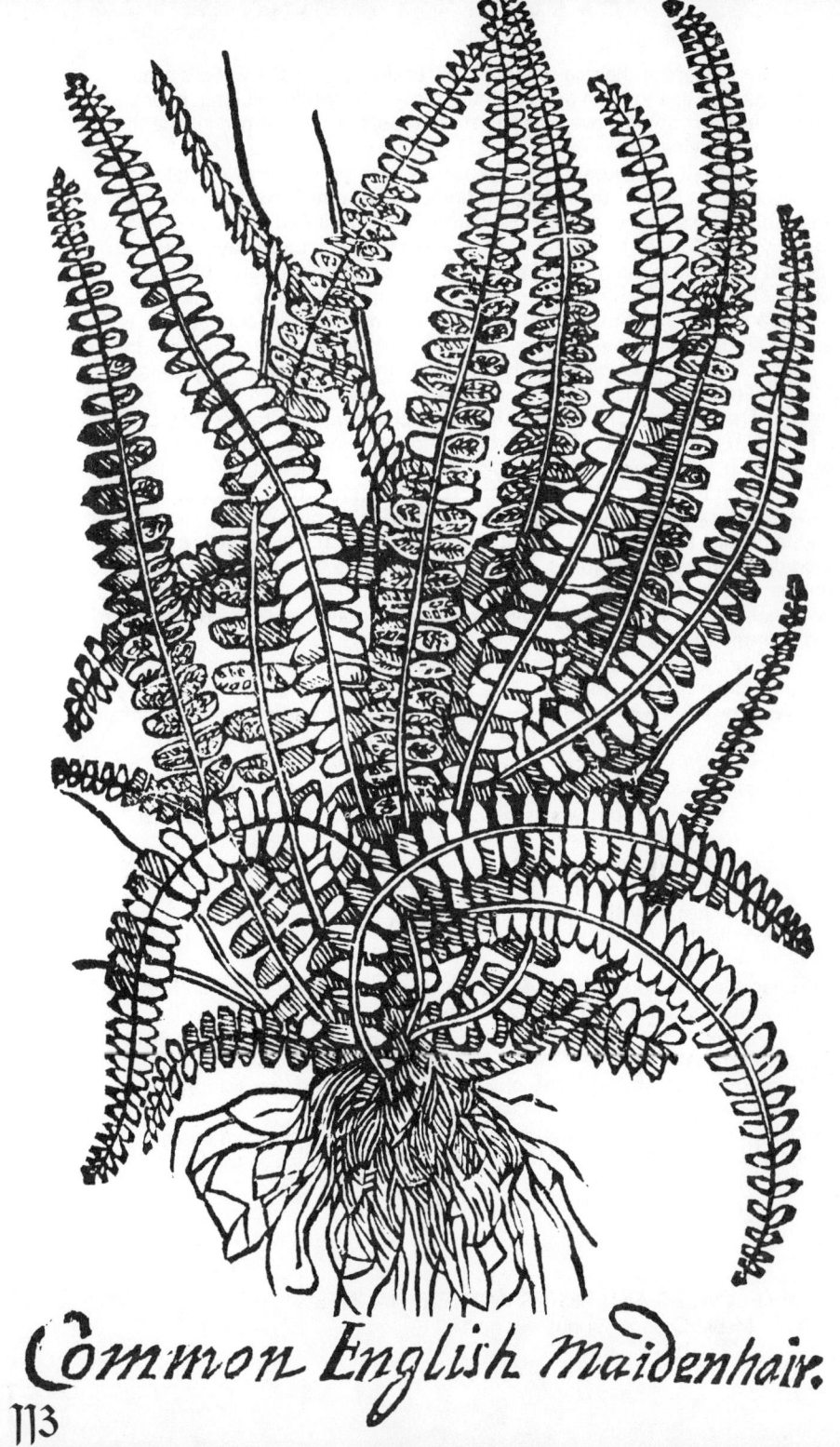

Common English Maidenhair.

In the old flower language, the Fern was the symbol of sincerity. In the wood language, the mystic speech of the Magician, the Fern stands for silence. Are not these interpretations the same?

The Fern is a voiceless sentinel of the silent woodlands; it has no flower to draw to it the hum of insects. Around the margins, or following the veins of its fronds, gather the intangible spores scarce deserving the name of seed till, in a further stage of development, they generate the dual forms which mutually perpetuate the race.

The Fern does not appeal directly to insect or man through a specialized color, or perfume. The wind passing through the trees of the forest, or among the reeds of the marshes, moves them to seeming articulate speech, but it tosses the heavily massed banks of Ferns, and sweeps the brake jungles on the wild commons, swaying them to and fro, while the silence that follows their motion is as deep as when the pad-footed cat hurries over soft turf, springs noiselessly, misses its quarry, and crouches once more,—to the eye a bewilderment of unheard action.

From the very circumstances of its evolution and growth, the Fern is more aloof than the flowering plants and also lacks the personal attributes which have given familiar names to blossoming things. These varied attributes have led flowers through the gates of poetry into the more serious realm of prose, until they not only have become a part of literature, but have a literature all their own, while their hold on household love increases like their race.

Not so with Ferns. They have scanty literature and few gracious names. Their tribal Golden Age had passed before man came to read their meaning. Back in the time of ancient life they were evolved, and held sway when fishes were the highest type of animals. Then gigantic forms of Ferns, Lycopods and Horsetails, did their work of absorbing the carbonic acid gas from the surcharged air, and transforming it into mighty forests, the only terrestial verdure. This work complete, the atmosphere purified, these forests were in their turn submerged, turned slowly to vast beds of coal, and higher plant forms appeared above them. Though the Fern tribe as a modified type remains, it has dwindled in numbers and stature until the extinct species far exceed the living, so that the tribe that once was all in all, now holds a little fiftieth part of the earth's flora, and is a mere background, as it were, for the varied forms, glowing colors, and soft perfumes which blend to dower the flowering plants with the fascination of personality.

"I wonder why Ferns are such nameless sort of things, not nearly so livable and lovable as flowers," said Flower Hat, as she leaned against a sloping rock, cushioned with moss and Polypody, cast aside her hat among a mass of Christmas Ferns, and rumpled her hair after a fashion of her own "to let it breathe," as she said, all the time fanning idly with a broad Fern frond.

It was the afternoon in early August when we had gone to Time o' year's woods, crossing Treebridge to find Rattlesnake Plantain and then to have a Fern hunt through haunts that were in part both moist and dry, continuing along the grassy meadow edges and strip of bog that, together with the river, bounded the woods on three sides. At that moment we sat resting, listening to the sound of the water coming down the rocky glen,—its voice deepened and strengthened by two days of steady rain,—and looking at the graceful draperies that the Ferns were casting about the rocks and trailing down the river banks, heaping their gauzy fabric so recklessly near the water's edge that it seemed as though a breeze would blow it in, while the long, pliant Lady Ferns, drooping, covered each other's roots until they had all the sinuous grace of vines.

"Of course it's because so few Ferns have easy rememberable English names, and the lack of the name, I suppose, is because Ferns have no flowers with color and shape to suggest it," continued Flower Hat. "We used to go on 'botany walks' when I was at school near Hartford. In those days even, Ferns seemed such dumb plants; and, to my obtuse mind, there were only three kinds. One was Maidenhair, which is easy to remember because it is quite unlike anything else. Another, the Climbing Fern, with scalloped leaves, is almost all rooted out by this time—the kind that twists its stalk around the wood Goldenrods and weeds in moist places; the vine sometimes ends in a spray covered with rusty dust, looking like seaweed or leaves that had gone wrong. The third was the Walking Fern, which grew high up on rocky places; a Fern that we had to scramble on our hands and knees to find. And when we found it, every one cried 'Ah! Oh!', yet it wasn't much of a Fern after all, even though it had a reasonable name. It was merely a tuft of lengthened-out leaves, each one stretching as far as it could, then dipping down to root at the end, and start another plant, like a sort of vegetable measuring-worm. The seed dust, spores, or whatever you call them, were scattered zigzag over the underside of some of the leaves, for all the world like the caraway seeds on cookies. These three Ferns I could remember, but all the rest seemed alike to me, common Ferns.

"Lately, however, since fate has decided that I must live in the real country for more than half the year, and I've taken to following you 'thorough brake, thorough briar' like an obedient spaniel, I've noticed a great deal of expression in these same common Ferns. They seem to have little ways all their own, and meanings, too, if we could read them, nothing wonderful, nothing really grand like what the trees whisper to one, only something airy and mysterious, —scraps of songs without words which they think to themselves perhaps."

"'If trees are Nature's thoughts or dreams,
　　And witness how her great heart yearns,
Then she has only shown, it seems,
　　Her lightest fantasies in ferns,'"

I quoted, "and if you wish to see a score or more of these common Ferns in their haunts, and call each one by a name easy to remember, this is the season, for all Ferns have reached perfection now; and this is the place also, for here in a half-mile circle through Time o' Year's country, grow most of the familiar landscape Ferns which you would find if you tramped New England over.

"Oh, you are eager—forward, march! Take a few steps, stand by that great rock and look down. Is not this place in truth haunt of the Ferns?"

From FLOWERS AND FERNS IN THEIR HAUNTS
By　Mabel Osgood Wright
　　　1901

BRAKE AND COLIC ROOT

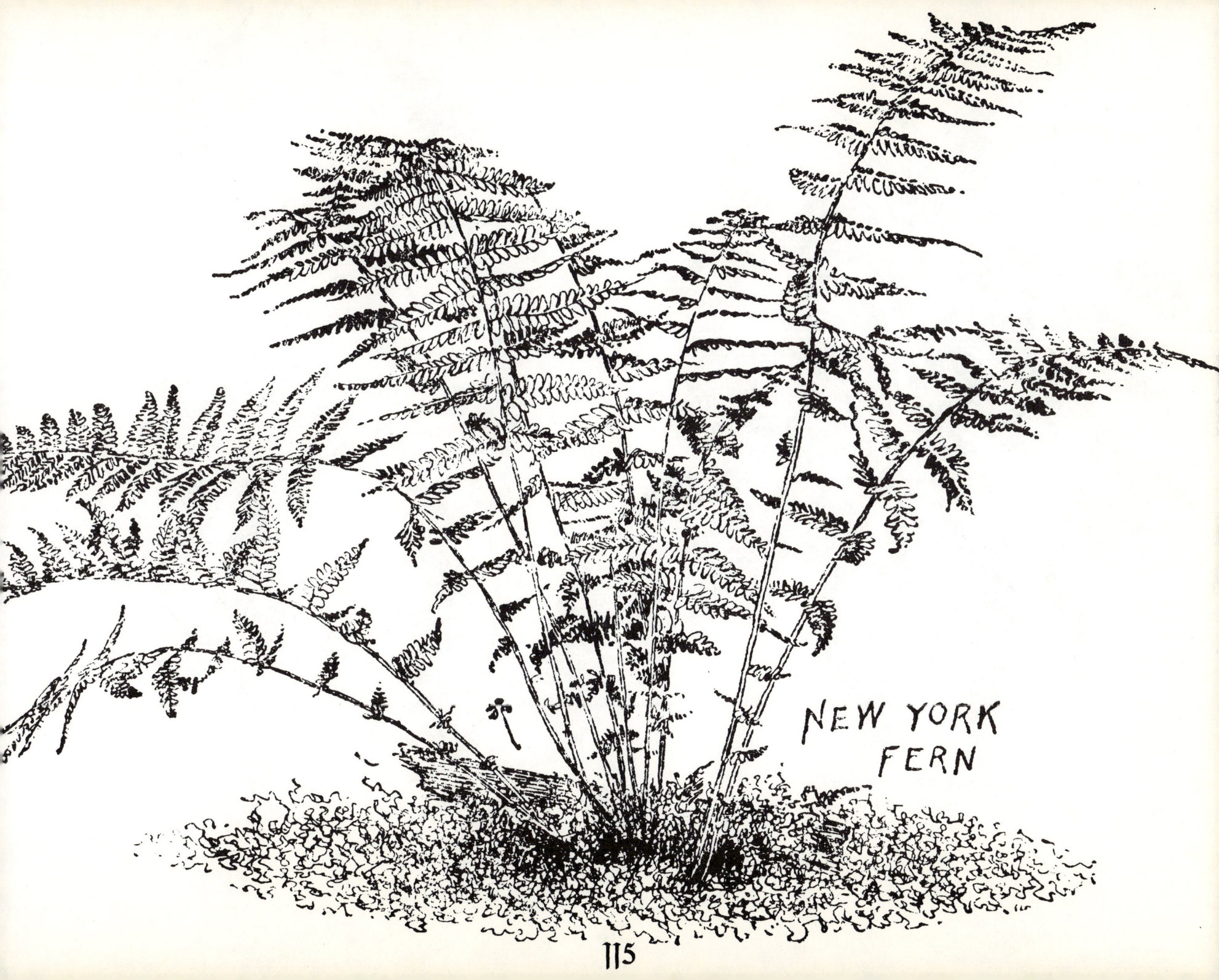

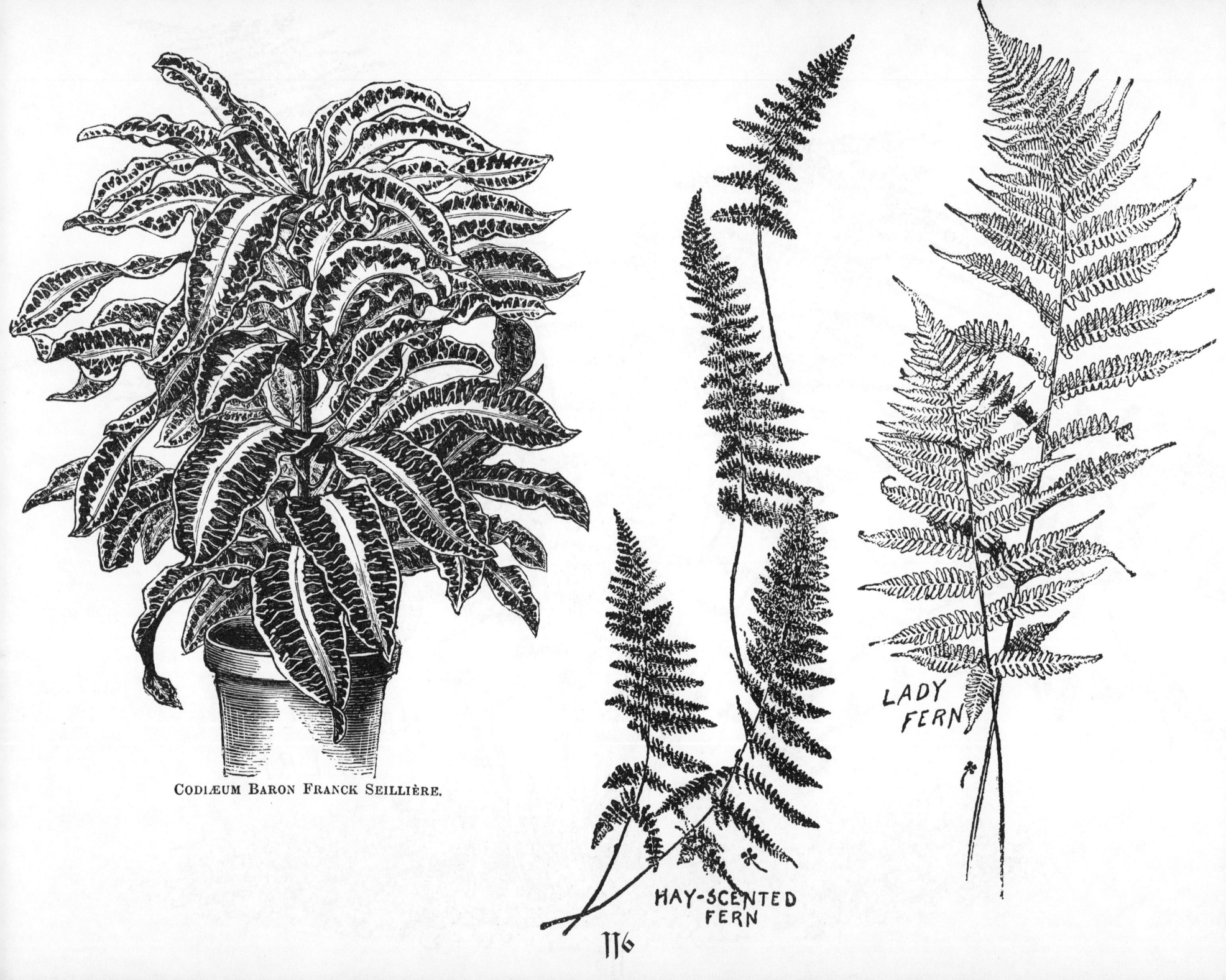

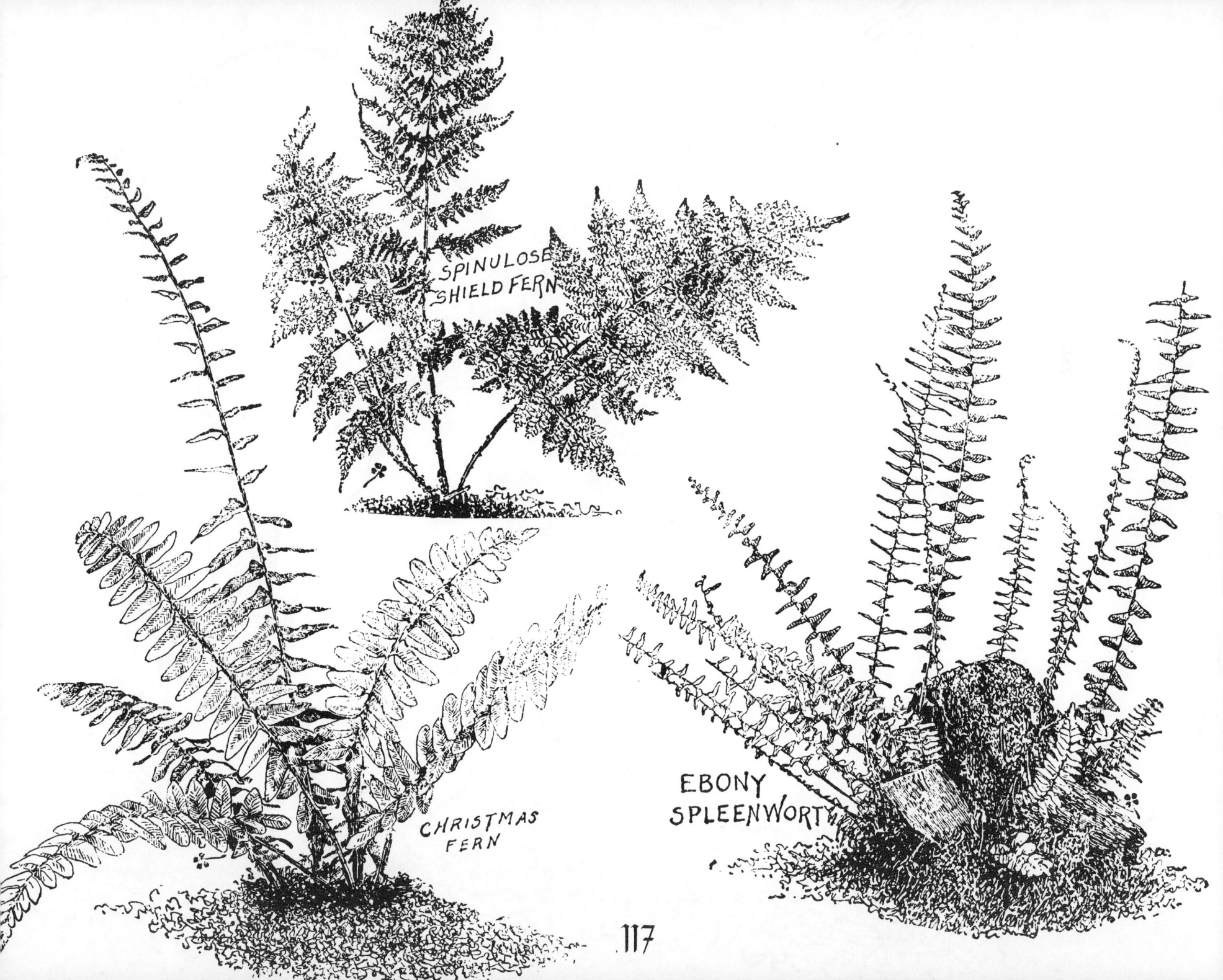

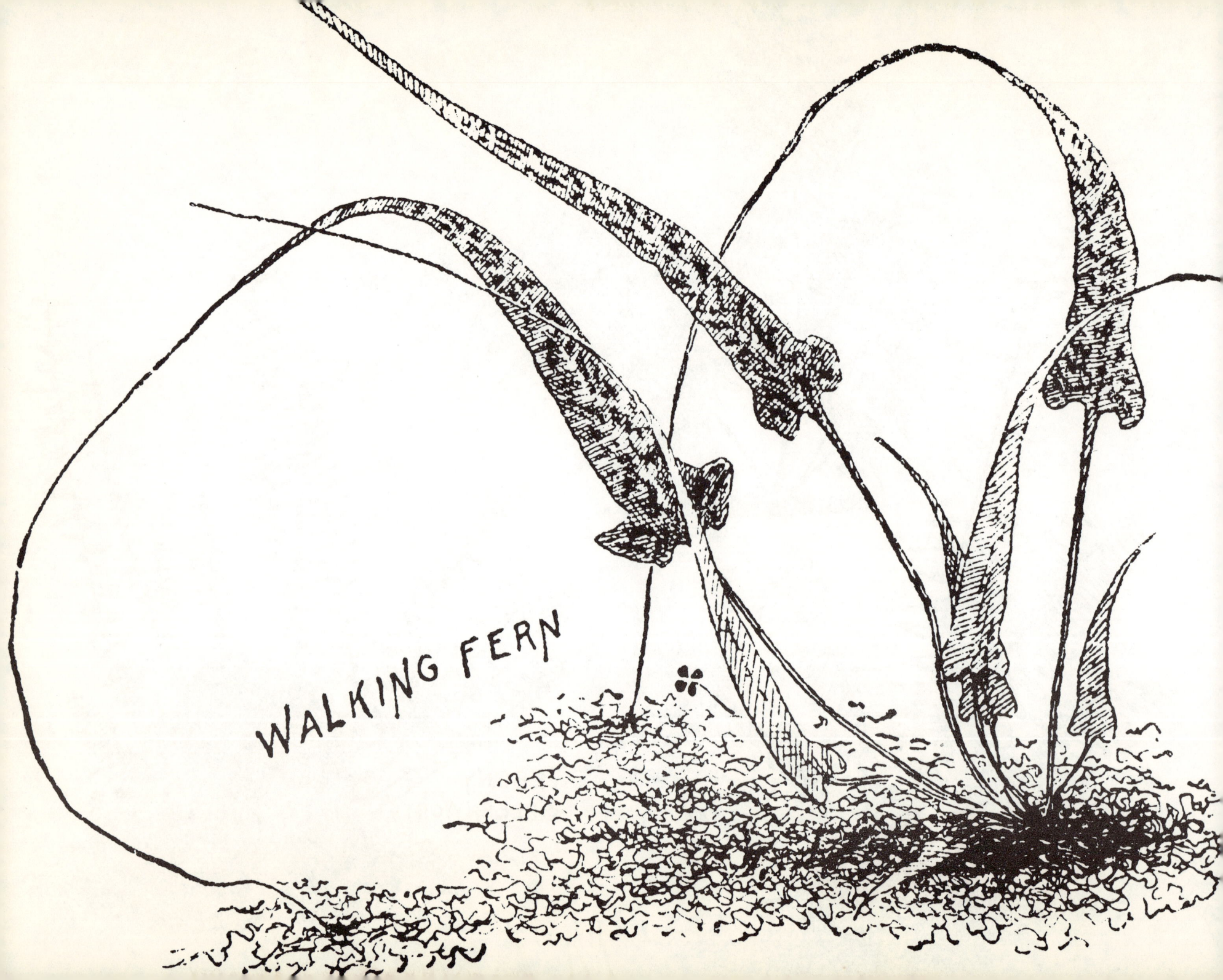

LADIES TRESSES

MEN & WOMEN IN THE GARDEN

ORNAMENTATION & DESIGN

"Furnished with whatever may make the place agreeable, melancholy, and country-like."
—Forest Trees, John Evelyn, 1670.

Quaint old books of garden designers show us that much more was contained in a garden two centuries ago, than now; it had many more adjuncts, more furnishings; a very full list of them has been given by Batty Langley in his NEW PRINCIPLES OF GARDENING, etc., 1728. Some seem amusing—as haystacks and woodpiles, which he terms "rural enrichments." Of water adornments there were to be purling streams, basins, canals, fountains, cascades, cold baths. There were to be aviaries, hare warrens, pheasant grounds, partridge grounds, dove-cotes, beehives, deer paddocks, sheep walks, cow pastures, and "manazeries" (menageries?); physic gardens, orchards, bowling greens, hop gardens, orangeries, melon grounds, vineyards, parterres, fruit yards, nurseries, sun-dials, obelisks, statues, cabinets, etc., decorated the garden walks. There were to be land gradings of mounts, winding valleys, dales, terraces, slopes, borders, open plains, labyrinths, wildernesses, "serpentine meanders," "rude-coppices," precipices, amphitheatres. His "serpentine meanders" had large opening spaces at proper distances, in one of which might be placed a small fruit garden, a "cone of evergreens," or a "Paradice-Stocks,"—about which latter mysterious garden adornment I think we must be content to remain in ignorance, since he certainly has given us ample variety to choose from without it.

Other "landscapists" placed in their gardens old ruins, misshapen rocks, and even dead trees, in order to look "natural."

In 1608 Henry Ballard brought out THE GARDENER'S LABYRINTH—a pretty good book, shut away from the most of us by being printed in black letter. He says:—

"The framing of sundry herbs delectable, with waies and allies artfully devised is an upright herbar."

Herbars, or arbors, were of two kinds: an upright arbor, which was merely a covered lean-to attached to a fence or wall; and a winding or "arch-arbor" standing alone. He names "archherbs," which are simply climbing vines to set "winding in arch-manner on withie poles." "Walker and sitters thereunder" are thereby comfortably protected from the heat of the sun. These upright arbors were in high favor; Ballard says they offered "fragrant savours, delectable sights, and sharpening of the memory."

Tree arbors were in use in Elizabethan times, platforms built in the branches of large trees. Parkinson called one that would hold fifty men, "the goodliest spectacle that ever his eyes beheld." A distinction was made between arbors and bowers. The arbor might be round or square, and was domed over the top: while the long arched way was a bower. In our Southern states that special use of the word bower is still universal, especially in the term Rose bowers. A quaint and universal furnishing of old Southern gardens were the trellises known as garden lyres.

Charming covered ways can be easily made by polling and training Plum or Willow trees. Arches are far too rare in American gardens. The few we have are generally old ones.

The many garden seats of the old English garden were perhaps its chief feature in distinction from American garden furnishings today. In a letter written from Kenilworth in 1575 the writer told of garden seats where he sat in the heat of summer, "feeling the pleasant whisking wynde." I have walked through many a large modern garden in the summer heat, and longed in vain for a shaded seat from which to regard for a few moments the garden treasures and feel the whisking wind, and would gladly have made use of the temporary presence of a wheelbarrow.

Seats of marble and stone are in many of our modern formal gardens.

Grottoes, arbors, and summer-houses were all of importance in those days, when in our latitude and climate men had not thought to build piazzas surrounding the house and shadowing all the ground floor rooms. We are beginning to think anew of the value of sunlight in the parlors and dining rooms of our summer homes, which for the past thirty or forty years have been so darkened.

Fountains were seen usually in handsome gardens; simple water jets falling in a handsome basin of marble or stone. Statuary of marble or lead was never common in old American gardens, though pretentious gardens had examples. Today, in our carefully thought-out gardens, the garden statuary is a thing of beauty and often of meaning. Usually our statues are of marble, sometimes a Japanese bronze is seen.

In the old black letter GARDENER'S LABYRINTH, a very full description is given of old modes of watering a garden. There was a primitive and very limited system of irrigation, the water being raised by "well-swipes"; there were very handy puncheons, or tubs on wheels, which could be trundled down the garden walk. There was also a formidable "Great Squirt of Tin," which was said to take "mighty strength" to handle, and which looked like a small cannon; with it was an ingenious bent tube of tin by which the water could be thrown in "great droppes" like a fountain. The author says of ordinary means of garden watering:—

"The common Watring Pot with us hath a narrow Neck, a Big Belly, Somewhat large Bottome, and full of little holes with a proper hole forced in the head to take in the water; which filled full and the Thumbe laid on the hole to keep in the aire may in such wise be carried in handsome Manner."

Beehives were once found in every garden; beeskepes they were called when made of straw. Picturesque and homely were the old straw beehives, and still are they used in England; the old one shown in the chapter on sun-dials can scarcely be mated in America. They served as a conventional emblem of industry. They were made of welts or ropes of twisted straw, as were the heavy winnowing skepes once used for winnowing grain.

The sadly picturesque old superstition of "telling the bees" of a death in a family and hanging a bit of black cloth on the hives as a mourning-weed still is observed in some country communities. Whittier's poem on the subject is wonderfully "countrified" in atmosphere, using the word chore-girl, so seldom heard even in familiar speech today and never found in verse elsewhere than in this rustic poem.

A pretty and appropriate garden furnishing was the cove-cote. The possession of a dove-cote in England, and the rearing of pigeons, was free only to lords of the manor and noblemen. When the colonists came to America, many of them had never been permitted to keep pigeons. In Scotland persistent attempts at pigeon-raising by folks of humble station might be punished with death. The settlers must have revelled in the freedom of the new land, as well as in the plenty of pigeons, both wild and domestic. In old England the dove-cote was often bukIt close to the kitchen door, that squab and pigeon might be near the hand of the cook. Dove-cotes in America were often simple boxes or houses raised on stout posts.

KNOTS AND PARTERRES.

IN the laying out of a garden much may depend upon the shape and disposition of beds in a parterre, or the arrangements of paths in a grass-plot. The designs are of quite a simple form, and are all drawn to fit into a square shape, but they can be easily altered to fill oblong, octagonal, or circular spaces. As suggestions for rose gardens, or for laying out any small enclosed space, with perhaps a sundial or fountain in the centre, they are particularly suitable.

MAZES.

MAZES and labyrinths may be traced from very early days. In England they are mentioned in the thirteenth century, when we read how fair Rosamund met her fate, in the labyrinth which concealed her bower, at the hands of the jealous Queen Eleanor. In early days a maze consisted of low hedges of privet, box, or hyssop, and William Lawson, writing in 1618, says: "Mazes well framed, of a man's height, may perhaps make your friend wander in gathering of berries till he cannot recover himself without your helpe."

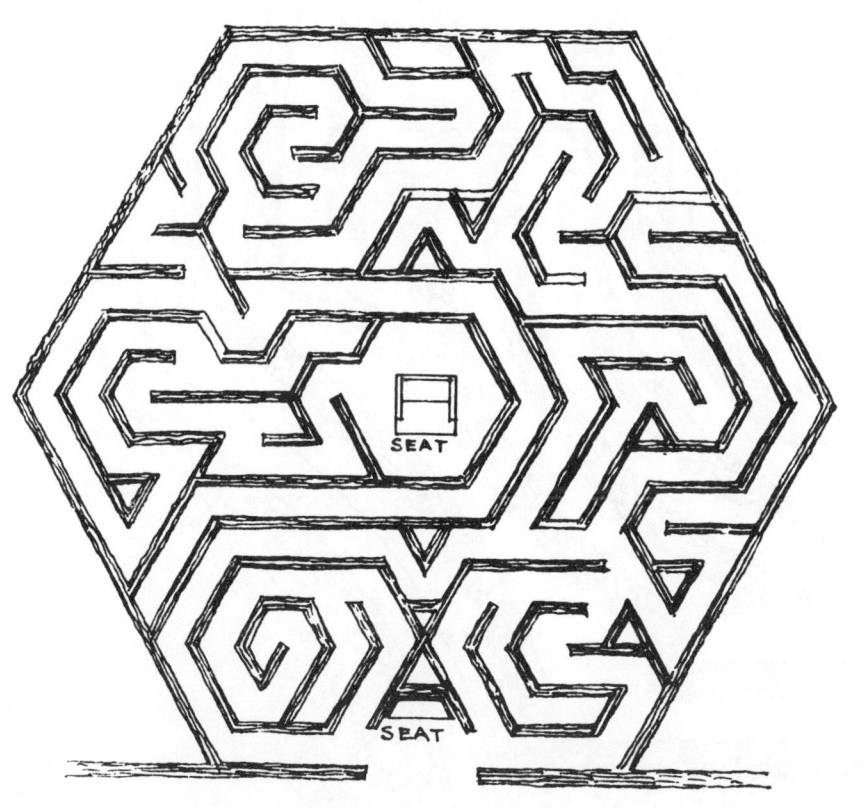

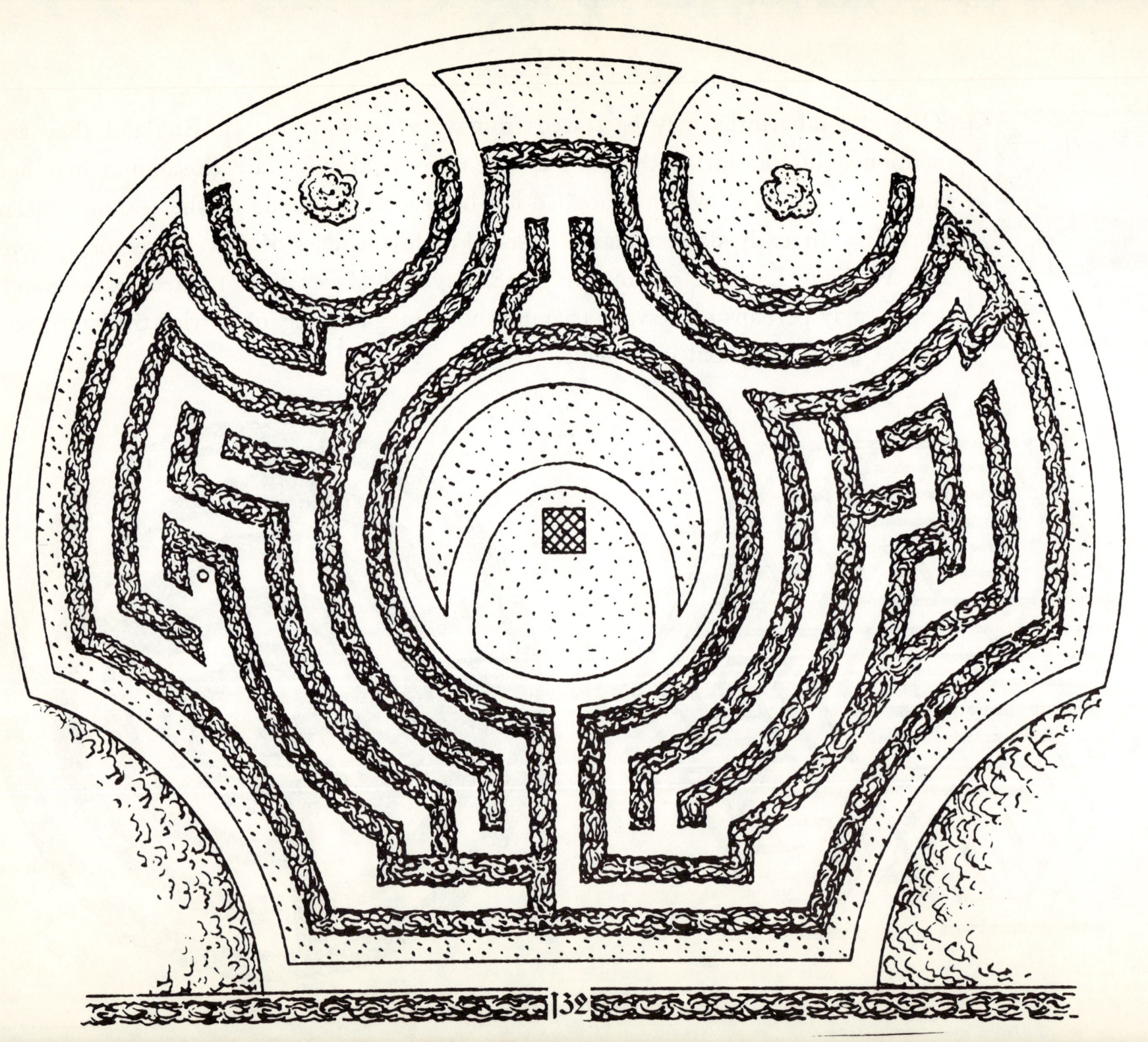

TOPIARY WORK

THE use of topiary work was first introduced into this country in the early years of the Tudor period, and soon growing into favour it became a conspicuous feature in gardens during the next two centuries. Although the system of cutting trees and shrubs eventually became very much abused, it serves to restrain the undue spreading and the amount of shadow caused by thick and heavy foliage, and there is no doubt that, rightly used, it invests a garden with attractive quaintness. Of the various trees most useful for topiary work, the yew is perhaps the best, and its rich green tones and soft velvety texture cannot be surpassed. Most of the examples left to us are of yew, partly owing to the fact that it is a slow grower, and, once having attained maturity, survives for many years. Privet, box, and rosemary were also used, but examples of topiary work in these are not frequently to be met with. The peacock was the form into which the trees were most commonly cut, and has always been a favourite device.

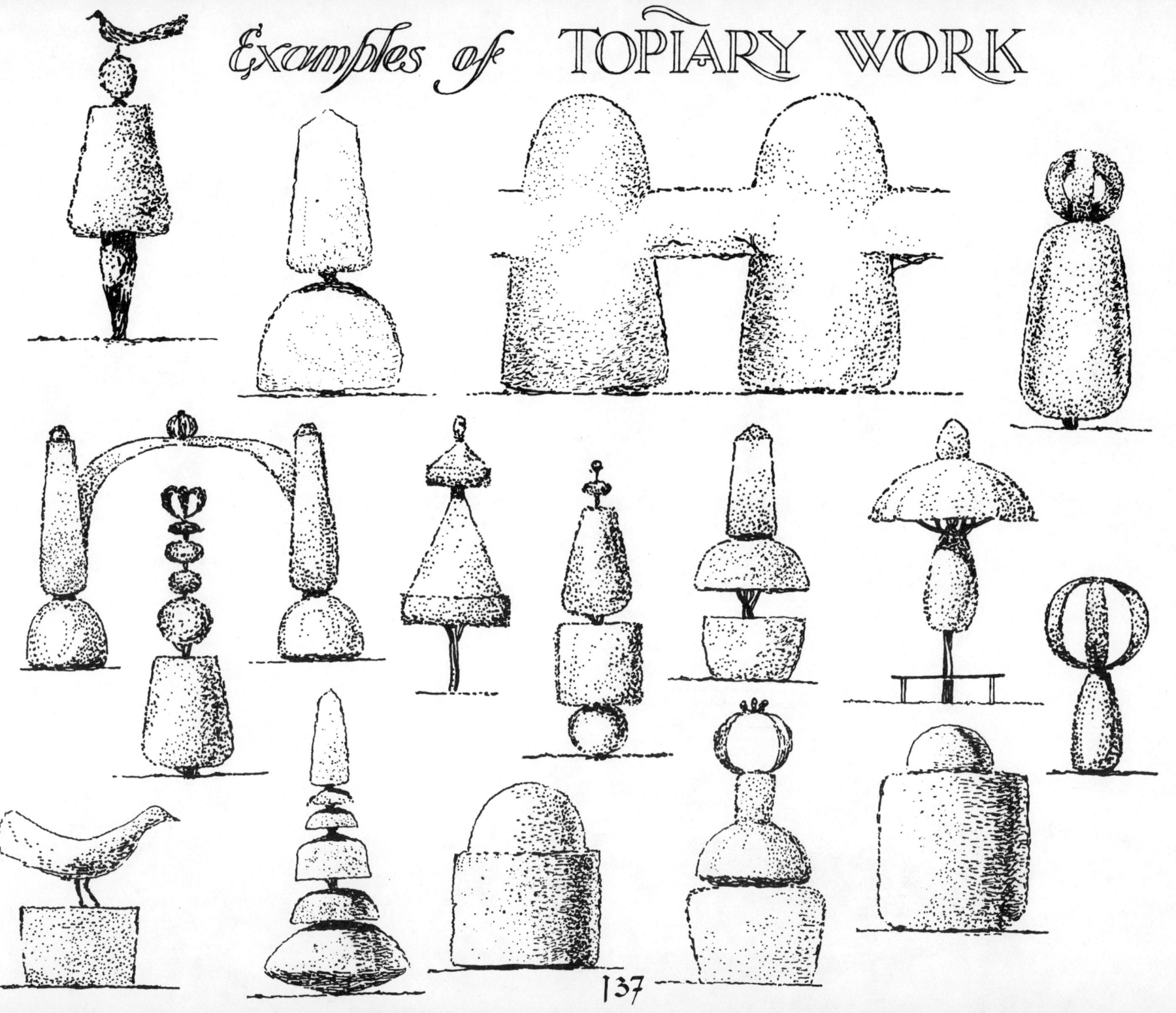

FOUNTAINS

A FOUNTAIN is perhaps the most delightful of all the ornamental accessories that go to complete a garden, and one in which the sculptor may find the greatest scope and freedom for his fancy and skill. On the Continent, especially in the garden schemes of Italy and France, fountains and other waterworks held a much more important position than in this country, where immense schemes such as those inaugurated by Le Notre at Versailles were never attempted.

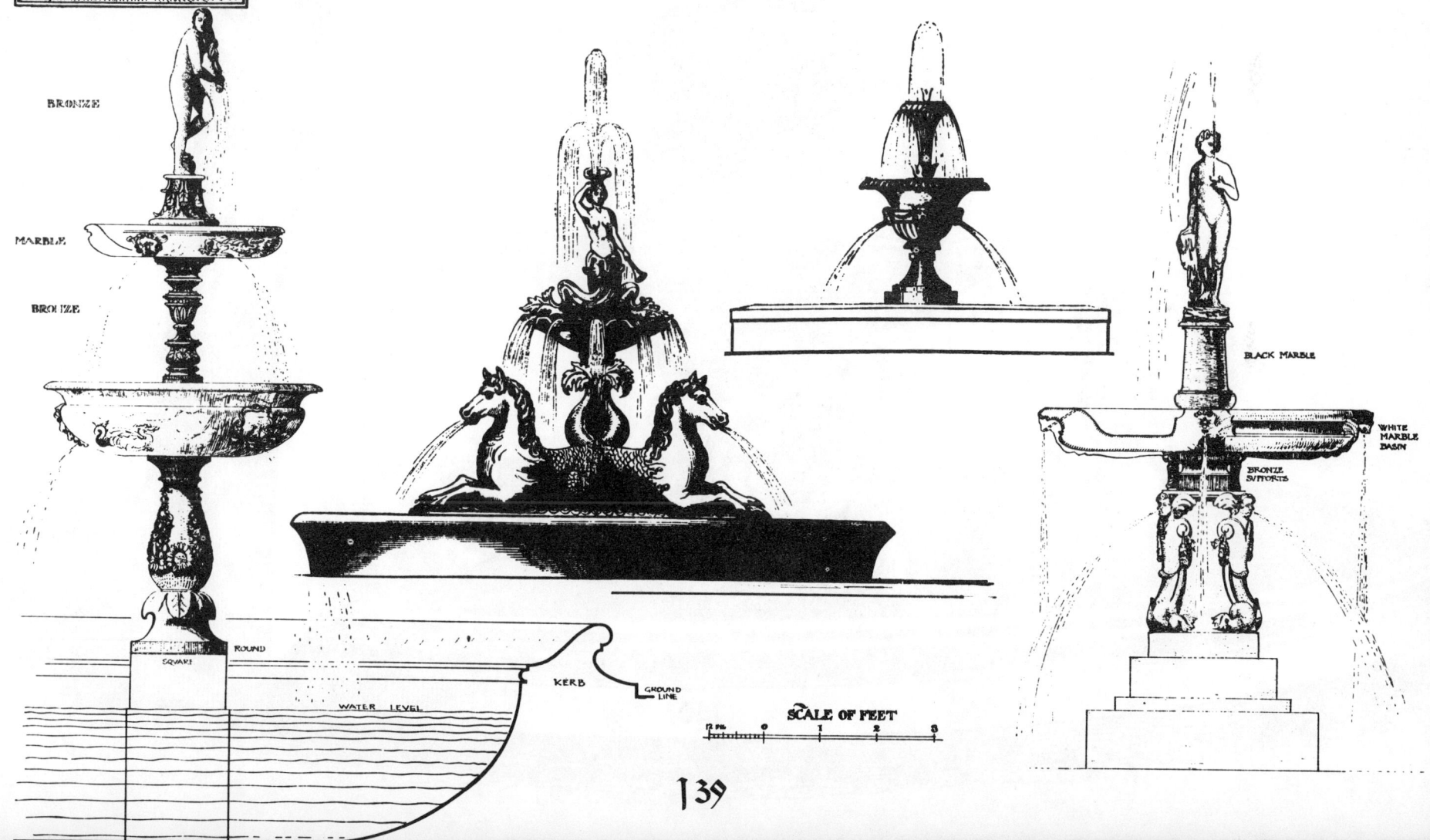

140

SPECIAL FOUNTAINS FOR SPECIAL GARDENS

SUNDIALS.

EVER since the sixteenth century sundials have occupied a foremost place among the ornamental adjuncts of a garden. Although, of course, they were originally regarded entirely from the utilitarian standpoint, it was not long before it became the custom to devote considerable attention and skill to their design, and they have frequently survived in their position when all other trace of the garden has disappeared. In these days it cannot be claimed that a sundial is of much practical use, yet everybody has a tender regard for them, and no formal garden would be considered complete without one. Their mottoes often serve to suggest the constant flow of time, or to inculcate a spirit of quietude and meditation. From the point of view of the garden designer, a sundial is often a very valuable accessory, as it may mark some prominent point, perhaps as the centre of a rose garden, or, when placed on a terrace, to lead the eye along some pleasant vista.

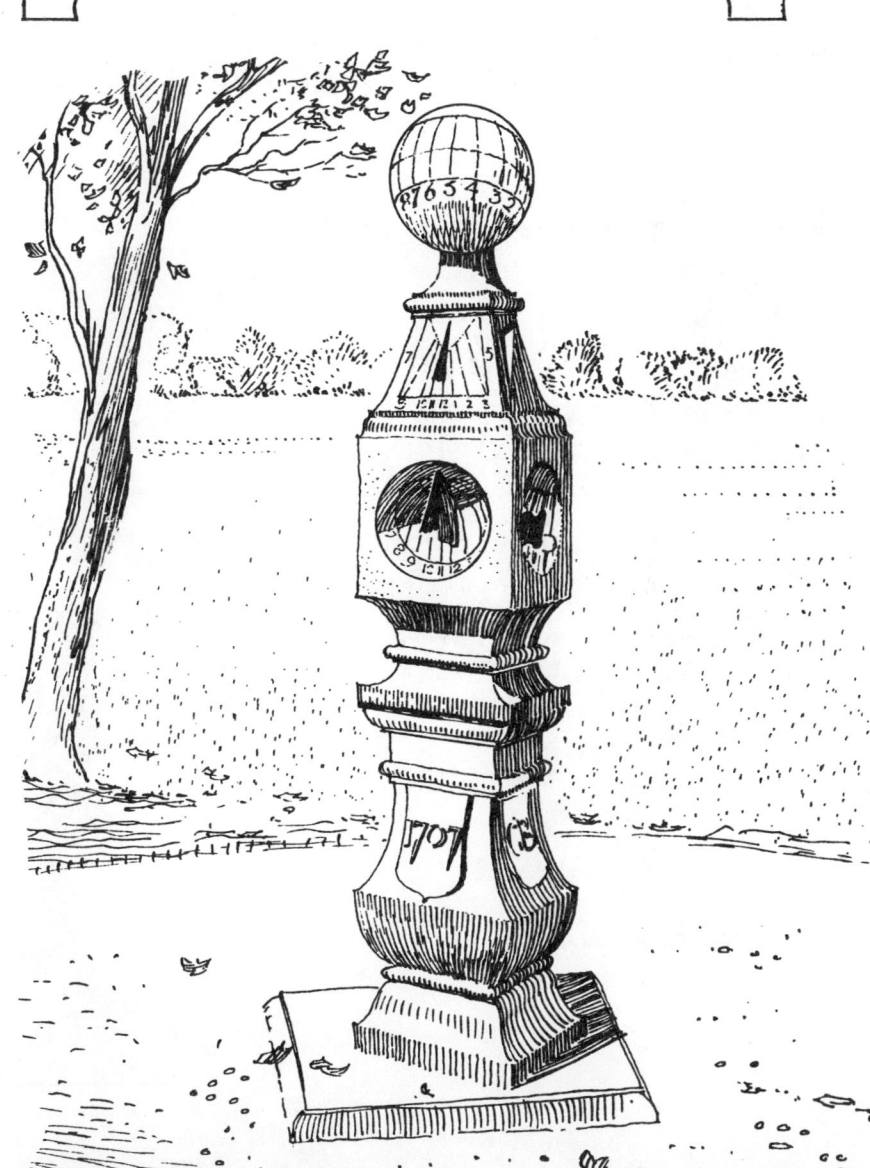

GARDEN LEADWORK.

AMONGST the many delightful accessories that go to make up the charm of a garden, few are more satisfactory than the figures, vases, and other objects formed of lead. The adaptability of this material and the delicacy of its colouring make it eminently suitable for such objects, and one can readily recall many instances of the fine effect produced by the soft silvery gray colour of a leaden figure against the rich green background of an old yew hedge.

145

TERRACES.

IN Tudor gardens terraces were usually placed in a position next to the enclosing walls of a garden, overlooking the surrounding country and forming a convenient point of vantage from which to view the arrangement of the garden plots.

GARDEN-HOUSES

THE Garden-house was the most important of all the accessories of a formal garden, from the early Tudor days to the middle of the eighteenth century, when it became the fashion to adorn gardens with Greek temples or Chinese pagodas, and the substantial and comfortable garden-house gave place to rustic wooden arbours overgrown with vegetation, which were far from being comfortable places from which to survey the beauty of the surroundings.

153

GATEPIERS.

IN nearly all old garden schemes, much attention was lavished on the gateways giving admittance to the various enclosures, and often, when every other vestige of the garden has disappeared, these remain, solitary survivors of the many details of all they enclosed and which surrounded them. Of the many different types of gatepier the most familiar is the square pier of brick or stone surmounted by a stone ball, either with or without a necking.

BENCHES, BRIDGES, SHADY SPOTS & WELLS

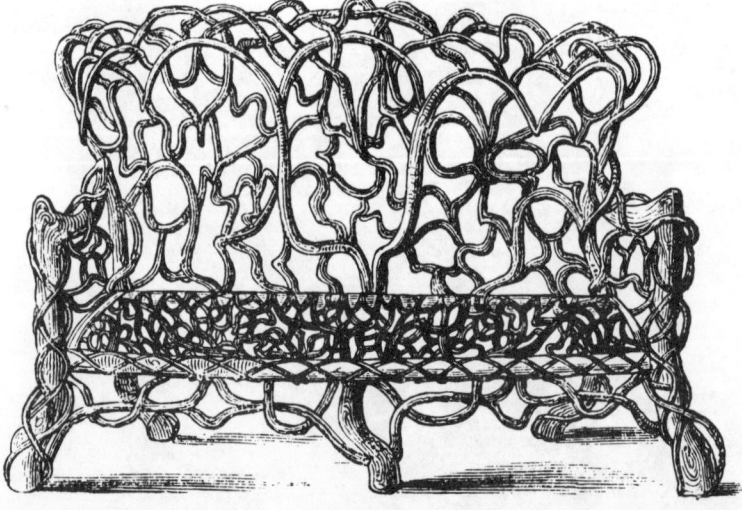

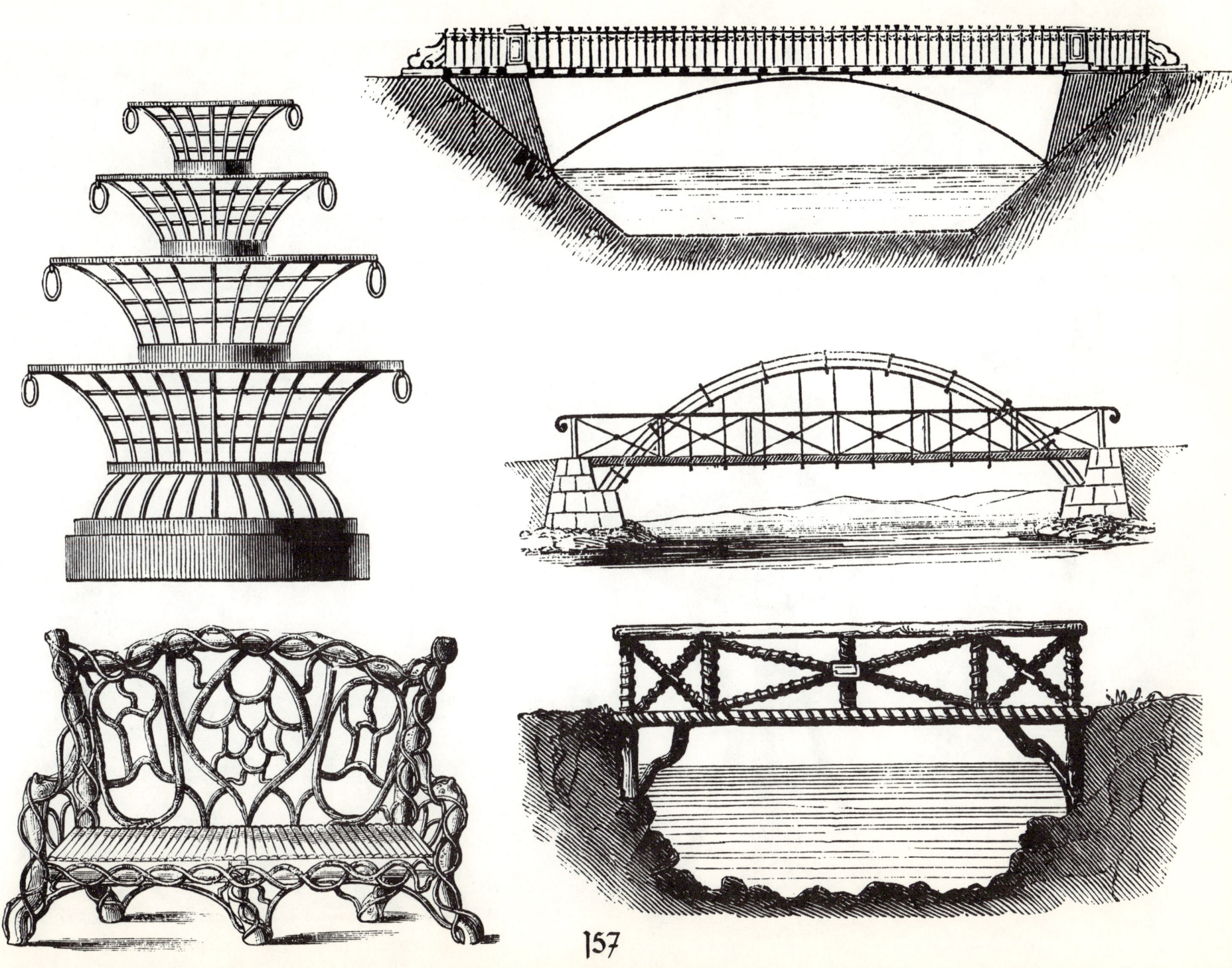

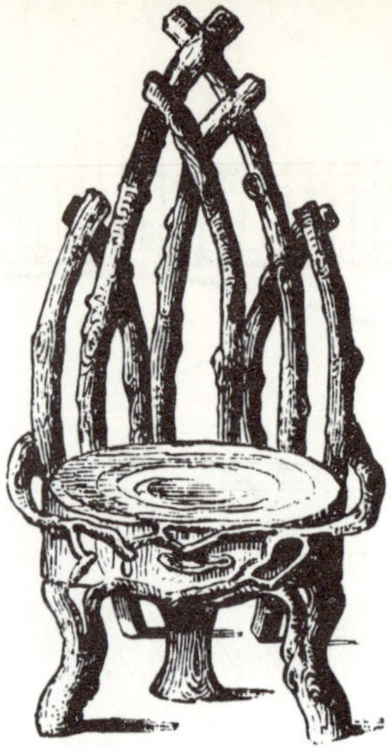

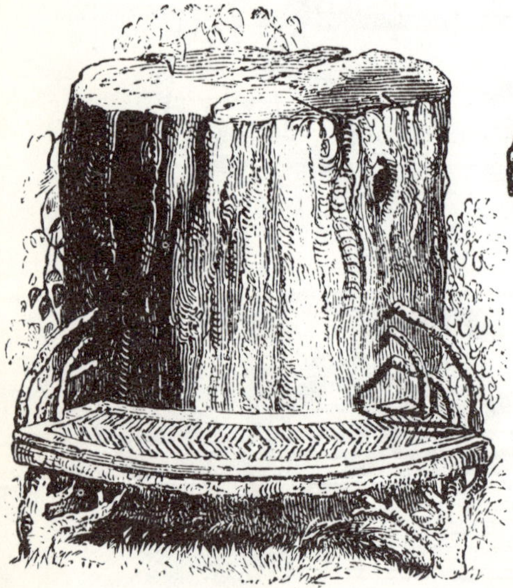

158

FLORA SYMBOLICA

FLORA SYMBOLICA

Shakespeare tells us that "fairies use flowers for their charactery," and so, he might have added, do mortals, for the language of flowers is almost as ancient and universal a one as that of speech.

The Chinese, whose chronicles antedate the historic records of all other nations, have, and ever seem to have had, a simple but complete mode of communicating ideas by means of florigraphic signs. The indestructible monuments of the mighty Assyrian and Egyptian races bear upon their venerable surfaces a code of floral telegraphy that Time has been powerless to efface, but whose hieroglyphical meaning is veiled, or, at the best but dimly guessed at in our day. India, whose civilization had attained its full vigor whilst that of Greece was in its cradle, has ever been poetically ingenious in finding in her magnificent Flora significations applicable to human interest. Biblical lore abounds in comparisons between "the golden stars that in earth's firmament do shine," and the feelings and passions of poor mortality. Persian poetry is replete with blossomy similes; whilst the mythology of the Greeks has been an apparently inexhustible storehouse to all authors in search of floral fancies. With the Hellenic race the symbolic language of flowers reached its culminating point of grandeur; and with the decline of Grecian glory faded away the brightest epoch in the history of florigraphy. In the eyes of the sterner-minded Latins this innocent study found less favor; and although they adapted many of their Hellenic predecessors' legends and customs, in connection with this science of sweet things, to their own mythology, yet so weakened was its hold upon the minds of the people, that when, in the course of events, the decadence of the Roman empire arrived, the attractive art was allowed to fade into comparative oblivion. With the revival of learning in the middle ages, this symbolic mode of correspondence once more re-bloomed, and, under the especial protection of chivalry, played a far from unimportant part in contemporary history. No gallant knight or gentle dame could then aspire to good breeding, unless perfectly conversant with florigraphy, as then taught; and the names, at least, of many of Europe's proudest families owe their origin to some circumstance connected with their founders' favorite blossom. In those days, minstrels and minnesingers sang praises of their mistresses' chosen bloom; the noblest knight and gayest squire broke many a lance, and emptied many a flagon, in honor of a sprig of broom, or a bunch of violets, that some fair dame had perchance adopted as her device. Even kings, not contended with their regal crowns, did not deem it derogatory to their dignity to enter the lists, in order to do battle for the floral wreaths that beauty proffered as a guerdon for the victor.

Thus every age and every clime promulgated its own peculiar system of floral signs; and although now-a-days, as regards the larger portion of Europe, the language is in many respects a dead one, yet still, amongst several Oriental races, this emblematic style of communication flourishes with much of its pristine importance.

"In Eastern lands they talk in flowers,
 And they tell in a garland their loves and cares;
Each blossom that blooms in their garden bowers
 On its leaves a mystical language wears."

It has been said that the language of flowers is as old as the days of Adam, and that the antiquity of floral emblems dates from the first throbbing of love in the human breast; and, indeed, to gain a glimpse of florigraphic symbolism, as it appeared in its earliest and freshest vigor, we should have to journey backwards far into the shadowy obscurity which envelopes the antediluvian history of mankind.

The love of flowers is felt and acknowledged by everybody, and in every land: it is a theme for every one, a feeling in which all can coincide. So universal a feeling—a feeling doubtless coeval with man's existence upon this globe—could not fail to be taken advantage of, and made subservient to, the passions of mortality; and to us it appears the most natural thing in the world that flowers should have been made emblematic and communicative agents of our ideas.

Florigraphy is a science that requires but little study. Some flowers, indeed, almost bear written upon their upturned faces the thoughts of which they are living representatives. That the "white investments" of the childlike Daisy should, as Shakespeare says, "figure innocence," is self-evident; that all nations should select the glowing Rose as an emblem of love could not be wondered at; whilst the little blue petals of the Myosotis palustris require no augur to explain their common name of Forget-me-not. Or again, who can doubt that the rich perfumes of some plants, or the sparkling lustres of others, must be deemed typical of joy and gladness; or that the melancholy hue and sombre looks of others should cause them to be selected to symbolize sadness and despair?

Simple as is the language of those bright earth stars, "the alphabet of the angels," as they have been somewhat inaptly styled, a great deal of skill may be expended in forming them into sentences, and much ingenuity may be exercised in explaining fully and satisfactorily the sentiments intended to be expressed towards the recipient of the floral message. The meanings of single token-flowers may seen be learned, but the knowledge of how to form them into a complete epistle does demand some little method. Many who use this fascinating style of correspondence frequently agree to adopt certain secret and original significations known only to themselves; and, if a little dexterity is shown, they not only give variety to, but also render their charming telegraphy perfectly unintelligible to the uniniated, although he may be the most skilled florigraphist breathing.

Every professor, ay, every student of this gentle art, may introduce new and varied combinations into its simple laws; but there are a few rudimentary rules that should not be neglected. An adept in the grammar of this language gives these directions of his pupils: "When a flower is presented in its natural position, the sentiment is to be understood affirmatively; when reversed, negatively. For instance, a rose-bud, with

its leaves and thorns, indicate fear with hope; but if reversed, it must be construed as saying, "you may neither fear nor hope." Again, divest the same rose-bud of its thorns, and its permits the most sanguine hope; deprive it of its petals and retain the thorns, and the worst fears may be entertained. The expression of every flower may be thus varied by varying its state or position. The Marigold is emblematic of pain: place it on the head, and it signifies trouble of mind; on the heart, the pangs of love; on the bosom, the disgusts of ennui. The pronoun I is expressed by inclining the symbol to the right, and the pronoun thou by inclining it to the left.

The richly varied and magnificent Flora of the American continent has offered her sons and daughters a floral vocabulary capable of almost unlimited application, and readily have the denizens of the New World seized upon the resuscitated these decaying systems of the Eastern Hemisphere. Many blossoms gathered from Columbia's well-stored garden will be discerned in this bouquet; but this bright bud of Charles Fenno Hoffmann's will not fail to increase the brilliancy of the tout ensemble:

THE LANGUAGE OF FLOWERS

"Teach thee their language? Sweet, I know no tongue,
   No mystic art those gentle things declare;
I ne'er could trace the schoolman's trick among
   Created things so delicate and rare.
Their language? Prithee! why, they are themselves
   But bright thoughts syllabled to shape and hue,—
The tongue that erst was spoken by the elves,
   When tenderness as yet within the world was new.

"And, oh! do not their soft and starry eyes—
   Now bent to earth, to heaven now meekly pleading,
Their incense fainting as it seeks the skies,
   Yet still from earth with freshening hope receding—
Say, do not these to every heart declare,
   With all the silent eloquence of truth,
The language that they speak is Nature's prayer,
   To give her back those spotless days of youth?"

Such are the tenets of florigraphists. Let us hope that such harmless if not beneficent doctrines are destined for universal acceptance, and that those bright times, foretold by Shelley, are not far distant, when
   "Not gold, not blood, the altar dowers,
   But votive blooms and symbol flowers."

HERB SYMBOLICA

ANGELICA Inspiration   Native to Eurasia
   Folklore says it was first seen in a dream by an angel as a cure for the plague.

BASIL Hatred   Introduced from India in 1548
   The Greeks and Romans thought basil would not grow unless you cursed it as you sowed the seed. There's a French idiom semer le basilic, which indicates slander or abuse.

BORAGE Courage or bluntness   Southern Europe and Northern Africa
It's said to instill courage, lighten the soul, drive away sadness.

BURNET Healing Herb   Eurasia
   Used as a tea to take down fevers, and heal wounds.

CHAMOMILE Energy in Action   Eurasia
   It is believed that chamomile will revive dying plants when it is planted next to them. It is a famous and soothing tea, said to restore a man to his senses.

CARAWAY   Eurasia
   A culinary herb, used occasionally in perfume.

CHIVES Small Burn   Eurasia
   The genus is Allium, and allium is from the Celtic meaning hot, or burning. Chives are one of the smallest and least hot of this onion family.

CORIANDER Hidden Merit   Southern Europe
   As seeds dry they become more and more fragrant. Seeds have been found in Egyptian tombs; it may be the manna of the Bible.

DILL Meeting Seed   Spain—grown in English gardens for over 300 years
   In England used to put babies to sleep, but in America eaten during the Sunday sermon, to keep awake! Hence Meeting Seed.

SWEET FENNEL Perception   Europe & Asia
   Thought to cure blindness, cataracts. Give you confidence and strength. In general, clarify your view of things.

LAVENDER Distrust   Mediterranean region
   Lavender oil if used in large amounts is a narcotic poison; in smaller amounts it is delightful and refreshing. Hence the symbolica may refer to knowledge of where to draw the line.

LEMON BALM Sympathy   Eurasia
   Meaning sympatico, making you loving and beloved. Beehives were also rubbed with it to attract new colonies.

LOVAGE Herb of the Sun   Southern Europe
   This plant was used to relieve flatulency, also means being wind-blown, pretentious.

Borrage.

SWEET MARJORAM Blushes   Europe & Eurasia
   An ancient love herb. Venus supposedly used it to heal a wound from Cupid's arrow. But it only made matters worse and she straightaway fell in love with Adonis. Perhaps that's why Romans put it in bridal wreaths.

MINT Virtue   Eurasia, Europe, and Australia
   Traditionally used as an air freshener. Has a clean smell and virtue ranks as clean.

NASTURTIUM Patriotism   South America
   Because its leaves resemble armor and the flowers an empty helmet. Patriotism equals soldier in this case.

PARSLEY Festivity   Sardinia—Introduced to England in 1548
   Greeks and Romans used it in their festive garlands, on account of its staying green so long.

PENNYROYAL Flee Away Eurasia
   Used as an insect repellent and in an amulet worn to ward off dizzines and faulty thinking. As an insect repellent it was often used against fleas.

ROSE GERANIUM Preference   South Africa
   Perhaps because it's preferred due to its rosiness.

ROSEMARY Remembrance   Mediterranean Region
   Remembrance meaning friendship. One legend says: As the Holy Family was fleeing Egypt, all the other bushes made noise as they brushed by them, letting their enemies hear. Only the Rosemary was silent, letting them pass safely. Quiet, deep friendship.

RUE Disdain   Eurasia and the Canary Islands
   It was given to Ulysses by Mercury to help him stand against the lure of Circe.

SAGE Health   Mediterranean region
   Used by Europeans and later by Asians as a rejuvenator. Sage tea said to lengthen life, stop grey hair (and reverse it).

SORREL Parental Affection   Europe and Asia
   Folklore has it that sorrel is born of patriots' blood, the first new growth on the field after the battle.

SOUTHERNWOOD Jesting   Southern Europe
   Used as an aphrodisiac, and by young boys, who rubbed the ashes on their faces to grow whiskers.

TANSY I declare against you.   Europe
   Tansies were eaten at Easter by Christians in remembrance of the bitter herbs the Jews ate at Passover.

TARRAGON Little Dragon   Eurasia
   Little Dragon is the translation of the French Esdragon, and Tarragon is a corruption of that word.

THYME Activity   Southern Europe—Introduced to England in 1548
   It has long been used to attract bees (the honey is much favored) perhaps the symbolica refers to that. As in "busy as a bee."

WORMWOOD Absence   Europe
   An allusion to Absinthe, the liqueur that employs wormwood. Refers to diminished brain activity.

YARROW Cure for Heartache   Eurasia
   An important medicinal herb. It was primary herb in an ointment made in castles and monastaries, for wounded soldiers coming home. Used as a tea for rheumatism and fevers. It takes its botanical name Achill ea in honor of Achilles, who was the first to use it on his wounds.

Dill.

SYMBOLICA EVERGREEN SHRUBS

LAUREL Glory   Mediterranean region
   The Greeks and Romans used a laurel crown as an emblem of all grand deeds. It was used as a medicine, spice, to ward off witchcraft and lightning, and clear the air of contagious diseases. "Crowning glory" must refer to this shrub. It was worn by warriors, poets, orators, philosophers, sovereigns, priests and priestesses.

ANDROMEDA "Will you help me?" Native to the bogs of Northeastern North America
   Linnaeus named it after the daughter of Cepheus and Cassiope, whose exposutre at the water-side, and rescue from the sea-monster by Perseus, is an episode in the fourth book of Ovid's "Metamorphoses." Linnaeus wrote: 'The plant is always fixed in some turfy hillock in the midst of the rock in the sea. . . As the distressed virgin cast down her blushing face through excessive affliction, so does the rosy-colored flower hang its head, growing paler and paler till it withers away. At length comes Perseus, in the shape of summer, dries up the surrounding waters, and destroys the monster."

CAMELLIA JAPONICA Supreme loveliness   China & Japan—introduced into Europe in 1639.
   Camellia was considered the rose of Japan by Europeans. The loveliest flower in the Japanese garden.

HEATH Solitude   Europe and N. America
   Because when you were feeling depressed or confused or anxious, "of diseased mind," the treatment was to get off by yourself, "forsake the common herd," and commune with Nature, out there, among the heath.

AZALEA Romance North temperate zone, chiefly North America and Eastern Asia.

RHODODENDRONS Agitation   Chiefly north temperate zone
   Rhododendron literally means Rosetree.

ANNUALS AND PERENNIALS

ACANTHUS   The Arts   Southern Europe
     A favorite of the Greeks and Romans as a form to copy   in architectural decoration. The principal decoration on the Corinthian Columns. 'Right well has this plant been chosen as an emblem of the arts. . . the more obstacles that are placed in its way, the more vigorously does it grow, and the more gracefully do its leaves curve, as if exalted and invigorated by the opposition which it encounters."

AMARANTH   Immortality   Tropical Old and New World   Most used species from West Indies, introduced to England around 1600.
     Because the flowers last so long. The Greeks called it "ever-lasting."

ANEMONE   Withered Hopes, Forsaken   North Temperate zone
     "Anemone was a symph beloved by Zephyr, and Flora, jealous of her beauty, banished her from her court, and finally transformed her into the flower that now bears her name." From anemos, the wind, inhabiting unprotected places; wind-flower.

ASPHODEL   "I will be faithful unto death."   Mediterranean region
     The Greeks dedicated it to the dead and planted it around graves. Homer said that you passed a plain of Asphodels after crossing the River Styx.

ASTER   After-thought   North America, Asia, and Europe
     "Because it begins to blow when other flowers are scarce. It is like an afterthought of Flora's, who smiles at leaving us."

BROOM Humility   Europe and Western Asia
     An emblem of Scotland to traveling Scots, who sang songs about this everyday and beautiful plant of their native land. Sort of a symbol of "There's no place like home."

BUTTERCUPS   Riches, Memories of Childhood   Europe and Southwestern Asia
     Chiefly an English symbol—refers to fields of buttercups that were common in rural Great Britain, nice to walk through, play in, easy for small fingers to pick.

CANTERBURY BELLS   "I will ever be constant."   Southern Europe
     Named after the town where they grow in profusion. They are the church bells of forest elves and fairies, that ring in alarm or joy depending, of course, on the situation. Constancy meaning they are relied on for their vigilance.

CAMPHIRE   Fragrance   N. Africa, Asia, and Australia
     Used as a perfume, very popular in Egypt. Women   perfumed themselves and their houses with the scent.

CELANDINE   Joys to come   Eurasia
     The swallow herb, thought to be used by adult swallows to cure blindness in their young. Culpepper recommended it for restoring eyesight. Its botanic name Cheliodonium is from the Greek for swallow.

CLEMATIS Artifice   Europe, Asia, Africa, N. America, and Australia
"Because, some say, beggars, in order to excite pity, made false ulcers—which, however, sometimes produced real ones—in their flesh by means of its twigs."

CLOVER Promise   Europe, N. Africa, N. America
The national emblem of Ireland. St. Patrick explained the mysteries of the Trinity using a clover leaf. Patriotism meaning hope for the future.

CONVOLVULUS, MINOR Night   Southern Europe and Eastern Asia
Because the flowers resemble a night-time sky full of stars.

CORN Abundance   Americas and West Indies
In this case the symbolica refers to the ancient word corn as "the generic name applied to all grain suitable for food." Thus Ceres was the goddess of corn. The botanic name for plant corn is zea and that is Greek for cereal.

CORNFLOWER Delicacy   Europe
Its classic surname is Cyanus who was "a fair young devotee of Flora, who made garlands for public festivals out of various sorts of wild flowers...accompanying her pleasant labor by singing the songs of her beloved fatherland."

COWSLIP Youthful Beauty   Europe, Eurasia
Symbolica refers to the way it looks, "gracefulness, slender stem..."

CROCUS Cheerfulness   Mediterranean region to Southwestern Asia. Introduced to British gardens around 1600.
A plant with a lot of stories attached to it. It was valued as a medicine, dye, and spice. "The golden dye...was held in great regard by the Irish peasantry, and became at last quite an emblem of their oppressed nationality: as such the color was very obnoxious to the English Government, and, finally, they passed laws against its use."

DAFFODIL Deceit   Mostly European and North African
A Greek emblem, on account of its narcotic qualities, "it delights heaven and earth with its odor and beauty, yet, at the same time, it produces stupor and even death."

DAHLIA Pomp   Mexico and Guatemala
The symbolica refers to the way it looks. Stately, pompous, like a politician about to say something.

DAISY Innocence   Europe
Whilst Belides (a wood nymph) was dancing with her favorite suitor, Ephigeus, she attracted the attention of Vertumnus, the guardian deity of orchards, and in order to shelter her from his pursuit she was changed into Boellis, or the daisy.

DANDELION Oracle   Europe
Used as a clock by shepherds because "its blossoms open and close at certain regular hour...and as a barometer, by predicting, by means of its feathery tufts, calm or stormy weather."

EVENING PRIMROSE Silent Love   North and South America
"She loves to look the pale moon in the face, and often in the witching hours of deep midnight...you may see the hospitable plant surrounded by such insects as avoid the light of day...which resort to her for their nightly banquet."

FORGET-ME-NOT Forget-me-not   Europe, N. Asia
There's a German legend that says a knight and his bethrothed were walking on the banks of the Danube when she saw a bunch of these flowers floating downstream. She said she would like to have them, and the knight immediately jumped into the river. He retrieved the flowers but the weight of his armor made it impossible to remount the slippery bank. He flung the flowers to her, crying, "forget me not," and drowned.

FOXGLOVE Insincerity   Europe, N. Africa, W. Asia
Because it is poisonous.

FUCHSIA Taste   Tropical America and New Zealand
So called because one of the early propagators, a Mr. Lee in London, was made rich by his decision to propagate this newly arrived flower, "the reward of the taste, decision, skill, and perseverance of old Mr. Lee."

HELIOTROPE Devoted Attachment   Peru
Heliotrope comes from two Greek words meaning the sun and to turn, "because of its having been supposed to turn continually towards the sun, following his course round the horizon."

HYACINTH Game Play   Mediterranean region and South America
Hyacinthus was a beautiful boy, sone of a Spartan king and favorite of Apollo. Zephyrus, jealous of their attachment, turned the direction of a quoit which Apollo had just pitched at play, and it hit Hyacinthus and killed him. Apollo turned the boy's body into this flower.

IRIS Message   Europe and Asia
The fleur-de-lis flower. Because of its brilliant and diversified hues, resembling a rainbow, it was named after the messenger of the gods, Iris.

THE LILY Majesty   Europe, Asia, W.Africa, N. America
Jesus said "consider the lilies of the field" describing them more glorious in their unadorned simplicity than Solomon, when arrayed in his most gorgeous apparel.

LOTUS Eloquence   Europe, Eurasia, Asia
"The Egyptians...consecrated the lotus to the sun, their god of Eloquence, and represented the dawn of day by a youth seated upon its blossom."

MARVEL OF PERU Timidity   Peru
Because in its native climate it only opens its flower at night. Timidity meaning it's too shy to bloom in the light when people can see it. However, in cooler climates it blooms all day. It's the temperature rather than light that affects the flower.

MIGONETTE Your qualities surpass your charms. Mediterranean and Red Sea regions
It quickly became naturalized in England, where it grows easily in city alleys, courtyards, where other flowers can't survive.

Dandelion.

MISTLETOE Give Me a Kiss   Europe
   Mistletoe was one of the most sacred plants to the Druids. At the beginning of their year the Druids would go in solemn procession into the woods to find this parasite. Animals and sometimes human beings were sacrificed. The mistletoe was then cut up and distributed among the people as the super cure-all. The tradition of kissing under the mistletoe traces back to Scandinavian mythology. Balder, the counterpart of Ap0llo, dedicated it to his mother Friga, equivalent to Venus. He placed mistletoe entirely under his control to prevent it from ever being used against his mother. The idea of being under the protection of a god led to being able to kiss any maiden who was under it, sort of an anything-goes policy, with Venus as your guide. The connection to Christmas is because it falls about the same time as the Druid New Year.

MOONWORT Honesty   Eurasia
   Honesty because of the transparency of the seed-vessel.

MOTHERWORT Concealed Love   Eurasia, Europe and Asia
   "This emblem of concealed love does not bloom until the second year, but, like the pure passion which it typifies, only blooms once."

MYRTLE Love   Mediterranean Region and Western Asia
   The Greeks and Romans consecrated the myrtle to Venus. "It was under the name of Myrtilla that Venus was worshipped in Greece."

NARCISSUS Self love   Mostly European and North African
   A beautiful youth from Boetia, Narcissus, fell in love with his own reflection in a pond. The nympth Echo, got angry because he was too busy looking at himself to notice her, and she cast a spell on him that rooted him to the spot, forever in love with himself. He pined to death and was metamorphosed by the gods into the flower that bears his name.

PERIWINKLE Tender Recollections   Southern Europe and Western Asia
   Rousseau has an anecdote about this plant. One day, when walking with a friend, she suddenly exlaimed, "Here is the periwinkle yet in flower!" He gave it a passing glance and didn't see it again until thirty years later, while walking with another friend. "Having then begun to botanize a little, in looking among the bushes by the way, I uttered a cry of joy: 'Ah, there is the periwinkle!' and so it was." He then had a vivid recollection of every incident of a particular time of his life.

PIMPERNEL Change   England and Southern Europe
   Another flower that acts as a clock and barometer for shepherds. Said to open its flower at eight in the morning and close around noon. It also is said to close its flower when it's about to rain, which would leave you with two options, either it's noon or it's going to rain.

POLYANTHUS Confidence   Europe and Asia
   It's one of the first flowers of spring. Plucky little things that lead rather than follow.

POPPY Consolation and Oblivion   Most are from the temperate regions of Europe and Asia
   The Symbolica refers to the effect of the opium poppy, and is meant in this order: peace and then nothingness.

ROSE Love   Varieties we grow now were mostly from Asia, but there are roses native to Europe and North America
   The favorite of all flowers, the most grown, the most hybridized. And the densest in sentiment. The Greeks said the rose was originally white, and received its red color from blood drawn by a thorn from the foot of Venus, as she was running to help her adored Adonis. There's seemingly endless symbolica on roses, from different countries, for different colors, and continually added to through literature. Love is the most pervasive.

SENSITIVE PLANT Bashful love   Brazil
   A remarkable plant, it curls its leaves inward, closes when you touch it.

SNOWDROP Friend in adversity, hope   Southern Europe and Asia Minor
   The first flower of spring. It signals the beginning of the end of winter.

SPEEDWELL Fidelity   Europe, Eurasia, Asia Minor, but chiefly New Zealand
   Also called Veronica, after St. Veronica who gave her handkerchief to Christ to wipe the perspiration from his face as he carried the cross to the crucifixion, because the blossom was said to resemble that handkerchief.

STOCK Lasting Beauty   Southeastern Europe
   Grown around one's home or one's castle, as the case may be, as a promoter of lasting beauty, chiefly by and for woman. Whether it was used as some sort of elixer is not clear, but it is clear it was highly valued and carefully cultivated.

SWEET WILLIAM Finesse, dexterity   Eurasia
   Because of "the charming manner in which it arranges its variegated blossoms into bouquet-shaped clusters." Gerarde says it's used "to deck up gardens, the bosoms of the beautiful, garlands, and crowns for pleasure."

SCOTCH THISTLE Independence   Europe
   The national emblem of Scotia, probably adopted around the middle of the fifteenth century. It was used in a wine and also worn around the neck to dispel melancholy, or rather what was considered "superfluous melancholy." It is also a symbol of surliness because of the Scotch motto that appears under it on the emblem "Nobody molests me with impunity."

TUBERROSE Dangerous Pleasures   East Indies and Mexico
   Its white blossom is said to exude an exquisite perfume which is so powerful that to enjoy it without danger it is necessary to keep some distance from the plant. It was worn by Malayan women in their hair to inform a lover that the suit was pleasing.

TULIP A declaration of love   Persia, Armenia, and S. Europe
   When tulips were first introduced to Holland in 1634 they set off a mania that rivals a gold rush. All classes of people were infected with the desire to own a specimen of the tulip. Incomes were wiped out, homes wrecked, as much as 5,000 pounds for one bulb has been recorded. Haarlem turned into a sort of stock market, with speculations rampant. After about three years the bottom fell out of the market and the crash came. But of course the Dutch still love the flower and it is still one of their major industries. From all this comes the symbolica, "a declaration of love," giving a tulip to someone was certainly an act of love since they were held so dear.

VIOLET Modesty   Europe, Asia, and North America
   Ia, the daughter of Midas and the bethrothed of Atys, whom, they say, to conceal her love from Apollo, Diana transformed into a violet.

WALLFLOWER Fidelity in Misfortune   Southern France
   In the middle ages, troubadours and minstrels wore it to symbolize the unchangeableness of their affection.

THE VOCABULARY OF FLORA SYMBOLICA

Abecedary	Volubility
Abatina	Fickleness
Acacia	Friendship
Acacia, Rose or White	Elegance
Acacia, Yellow	Secret love
Acanthus	The fine arts, Artifice
Acalia	Temperance
Achillea Millefolia	War
Achimenes Cupreata	Such worth is rare
Aconite (Wolfsbane)	Misanthropy
Aconite, Crowfoot	Lustre
Adonis, Flos	Sad memories
African Marigold	Vulgar minds
Agnus Castus	Coldness, Indifference
Agrimony	Thankfulness, Gratitude
Almond, Common	Stupidity, Indiscretion
Almond, Flowering	Hope
Almond, Laurel	Perfidy
Allspice	Compassion
Aloe	Grief, Religious superstition, Bitterness
Althaea Frutex (Syrian Mallow)	Persuasion
Alyssum, Sweet	Worth beyond beauty
Amaranth, Globe	Immortality, Unfading love
Amaranth (Cockscomb)	Foppery, Affection
Amaryllis	Pride, Timidity Splendid beauty
Ambrosia	Love returned
American Cowslip	Divine beauty
American Elm	Patriotism
American Linden	Matrimony
American Starwort	Welcome to a stranger, Cheerfulness in old age
Amethyst	Admiration
Andromeda	Self-sacrifice
Anemone (Zephyr Flr.)	Sickness, Expectation
Anemone, Garden	Forsaken
Angelica	Inspiration, or Magic
Apricot Blossom	Doubt
Apple	Temptation
Apple Blossom	Preference, Fame speaks him great and good
Apple, Thorn	Deceitful charms
Apocynum (Dogsbane)	Deceit
Arbor Vitae	Unchanging friendship, Live for me
Arbutus	Thee only do I love
Arum (Wake Robin)	Ardor, Zeal
Ash-leaved Trumpet Flower	Separation
Ash, Mountain	Prudence, or With me you are safe
Ash Tree	Grandeur
Aspen Tree	Lamentation, or Fear
Aster (China)	Variety, After-thought
Asphodel	My regrets follow you to the grave.
Auricula	Painting
Auricula, Scarlet	Avarice
Austurtium	Splendor
Autumnal Leaves	Melancholy
Azalea	Temperance
Bachelor's Button	Celibacy
Balm	Sympathy
Balm, Gentle	Pleasantry
Balm of Gilead	Cure, Relief
Balsam, Red	Touch me not, Impatient resolves
Balsam, Yellow	Impatience
Barberry	Sharpness of temper
Basil	Hatred
Bay Leaf	I change but in death
Bay (Rose) Rhododendron	Danger, Beware
Bay Tree	Glory
Bay Wreath	Reward of merit
Bearded Crepis	Protection
Beech Tree	Prosperity
Bee Orchis	Industry
Bee Ophrys	Error
Begonia	Deformity
Belladonna	Silence, Hush!
Bell Flower, Pyramidal	Constancy
Bell Flower (sm. white)	Gratitude
Belvedere	I declare against you
Betony	Surprise
Bilberry	Treachery
Bindweed, Great	Insinuation, Importunity
Bindweed, Small	Humility
Birch	Meekness
Birdsfoot (Trefoil)	Revenge
Bittersweet, Nightshade	Truth
Black Poplar	Courage, Affliction
Blackthorn	Difficulty
Bladder Nut Tree	Frivolity, Amusement
Bluebottle (Centaury)	Delicacy
Bluebell	Constancy, Sorrowful regret
Blue-flowered Greek Valerian	Rupture
Bonus Henricus	Goodness
Borage	Bluntness
Box Tree	Stoicism
Bramble	Lowliness, Envy, Remorse
Branch of Currants	You please all
Branch of Thorns	Severity, Rigor
Bridal Rose	Happy love
Broom	Humility, Neatness
Browallia Jamisonii	Could you bear poverty?
Buckbean	Calm repose
Bud of White Rose	A heart ignorant of love
Buglos	Falsehood
Bulrush	Indiscretion, Docility
Bundle of Reeds, with their Panicles	Music
Burdock	Importunity, Touch me not
Bur	Rudeness, You weary me
Buttercup (Kingcup)	Ingratitude, Childishness
Buttercups	Riches
Butterfly Orchis	Gaiety
Butterfly Weed	Let me go

Carnation, Striped	Refusal
Carnation, Yellow	Disdain
Cardinal Flower	Distinction
Catchfly	Snare
Catchfly, Red	Youthful love
Catchfly, White	Betrayed
Cattleya	Mature charms
Cattleya, Pineli	Matronly grace
Cedar	Strength
Cedar of Lebanon	Incorruptible
Cedar Leaf	I live for thee
Celandine, Lesser	Joys to come
Cereus, Creeping	Modest genius
Centaury	Delicacy
Champignon	Suspicion
Chequered Fritillary	Persecution
Cherry Tree, White	Good education
Cherry Blossom	Insincerity
Chestnut Tree	Do me justice
Chinese Primrose	Lasting love
Chickweed	Rendezvous
Chicory	Frugality
China Aster	Variety
China Aster, Double	I partake your sentiments

Cabbage	Profit
Cacalia	Adulation
Cactus	Warmth
Calla Aethiopica	Magnificent beauty
Calceolaria	I offer you pecuniary assistance, or I offer you my fortune
Calycanthus	Benevolence
Camellia Japonica, Red	Unpretending excellence
Ditto, White	Perfected loveliness
Camomile	Energy in adversity
Campanula Pyramida	Aspiring
Camphire	Fragrance
Canary Grass	Perseverance
Candytuft	Indifference
Canterbury Bell	Acknowledgment
Cape Jasmine	I am too happy
Cardamine	Paternal error
Carnation, Deep Red	Alas! for my poor heart

China Aster, Single	I will think of it
China or Indian Pink	Aversion
China Rose	Beauty always new
Chinese Chrysanthemum	Cheerfulness under adversity
Chorozema Varium	You have many lovers
Christmas Rose	Relieve my anxiety
Chrysanthemum, Red	I love
Chrysanthemum, White	Truth
Chrysanthemum, Yell.	Slighted love
Cineraria	Always delightful
Cinquefoil	Maternal affection
Circaea	Spell
Cistus, or Rock Rose	Popular favor
Cistus, Gum	I shall die tomorrow
Citron	Ill-natured beauty
Clarkia	The variety of your conversation delights me
Clematis	Mental beauty. Artifice
Clematis, Evergreen	Poverty
Clianthus	Worldliness. Self-seeking

Clotbur	Rudeness, Pertinacity
Cloves	Dignity
Clover, Four-leaved	Be mine
Clover, Red	Industry
Clover, White	Think of me, Promise
Cobaea	Gossip
Cockscomb Amaranth	Foppery, Affectation, Singularity
Colchicum, or Meadow Saffron	My best days are past
Coltsfoot	Justice shall be done
Columbine	Folly
Columbine, Purple	Resolved to win
Columbine, Red	Anxious and trembling.
Convolvulus	Bonds
Convolvulus, Bl.(Min.)	Repose, Night
Convolvulus, Major	Extinguished hopes
Convolvulus, Pink	Worth sustained by judicious and tender affection
Corchorus	Impatient of absence
Coreopsis	Always cheerful
Coreopsis Arkansa	Love at first sight
Coriander	Hidden worth
Corn	Riches
Corn, Broken	Quarrel
Corn Straw	Agreement
Corn Bottle	Delicacy
Corn Cockle	Gentility
Cornflower	Delicacy
Cornel Tree	Duration
Coronella	Success crown your wishes
Cosmelia Subra	The charm of a blush
Cowslip	Pensiveness, Winning grace, Youthful beauty
Cowslip, American	Divine beauty
Crab (Blossom)	Ill nature
Cranberry	Cure for heartache
Creeping Cereus	Horror
Cress	Stability, Power
Crocus	Abuse not, Impatience
Crocus, Spring	Youthful gladness
Crocus, Saffron	Mirth, Cheerfulness
Crown, Imperial	Majesty, Power
Crowshill	Envy
Crowfoot	Ingratitude
Crowfoot (Aconite-lvd.)	Luster
Cuckoo Plant	Ardor
Cudweed, American	Unceasing remembrance
Currant	Thy frown will kill me
Cuscuta	Meanness
Cyclamen	Diffidence
Cypress	Death. Mourning
Daffodil	Regard, Unrequited love
Dahlia	Instability, Pomp
Daisy	Innocence
Daisy, Garden	I share your sentiments
Daisy, Michaelmas	Farewell, or Afterthought

Daisy, Parti-colored	Beauty
Daisy, Wild	I will think of it
Damask Rose	Brilliant complexion
Dandelion	Rustic oracle

Dandelion, or Thistle-seed-head	Depart
Daphne	Glory, Immortality
Daphne Odora	Painting the lily
Darnel	Vice
Dead Leaves	Sadness
Deadly Nightshade	Falsehood
Dew Plant	A serenade
Dianthus	Make haste
Diosma	Your simple elegance charms me
Dipteracanthus Spectabilis	Fortitude
Diplademia Crassinoda	You are too bold
Dittany of Crete	Birth
Dittany of Crete, White	Passion
Dock	Patience
Dodder of Thyme	Baseness
Dogsbane	Deceit, Falsehood
Dogwood	Durability
Dragon Plant	Snare
Dragonwort	Horror
Dried Flax	Utility
Ebony tree	Blackness
Echites Atropurpurea	Be warned in time
Eglantine (Sweetbriar)	Poetry, I wound to heal
Elder	Zealousness
Elm	Dignity
Enchanter's Nightshade	Witchcraft, Sorcery
Endive	Frugality
Escholzia	Do not refuse me
Eupatorium	Delay
Evening Primrose	Silent Love
Ever-flowing Candytuft	Indifference
Evergreen Clematis	Poverty
Evergreen Thorn	Solace in adversity
Everlasting	Never-ceasing remembrance
Everlasting Pea	Lasting pleasure

Fennel	Worthy all praise, strength	Gooseberry	Anticipation
Fern	Fascination, Magic, Sincerity	Gourd	Extent, Bulk
		Grammanthus Chloraflora	Your temper is too hasty
Ficoides, Ice Plant	Your looks freeze me	Grape, Wild	Charity
Fig	Argument	Grass	Submission, Utility
Fig Marygold	Idleness	Guelder Rose	Winter, Age
Fig Tree	Prolific		
Filbert	Reconciliation	Handflower Tree	Warning
Fir	Time	Harebell	Submission, Grief
Fir Tree	Elevation	Hawkweed	Quick-sightedness
Flax	Domestic industry, Fate, I feel your kindness	Hawthorn	Hope
		Hazel	Reconciliation
Flax-lvd. Golden-locks	Tardiness	Heart's-ease, or Pansy	Thoughts
Fleur-de-Lys	Flame, I burn	Heath	Solitude
Fleur-de-Luce	Fire	Helenium	Tears
Flowering Fern	Reverie	Heliotrope	Devotion or, I turn to thee
Flowering Reed	Confidence in Heaven		
Flower-of-an-Hour	Delicate beauty	Hellebore	Scandal, Calumny
Fly Orchis	Error	Helmet Flower (Monkshood)	Knight-errantry
Flytrap	Deceit	Hemlock	You will be my death
Fool's Parsley	Silliness	Hemp	Fate
Forget-me-not	Forget-me-not	Henbane	Imperfection
Foxglove	Insincerity	Hepatica	Confidence
Foxtail Grass	Sporting	Hibiscus	Delicate beauty
Franciscea Latifolia	Beware of false friends	Holly	Foresight
French Honeysuckle	Rustic beauty	Holly Herb	Enchantment
French Marygold	Jealousy	Hollyhock	Ambition, Fecundity
French Willow	Bravery and humanity	Honesty	Honesty, Fascination
Frog Ophrys	Disgust	Honey Flower	Love sweet and secret
Fuller's Teasel	Misanthropy	Honeysuckle	Generous and devoted affection
Fumitory	Spleen		
Fuchsia, Scarlet	Taste	Honeysuckle, Coral	The color of my fate
Furze, or Gorse	Love for all seasons Anger	Honeysuckle, French	Rustic beauty
		Hop	Injustice
		Hornbeam	Ornament
Garden Anemone	Forsaken	Horse Chestnut	Luxury
Garden Chervil	Sincerity	Hortensia	You are cold
Garden Daisy	I partake your sentiments	Houseleek	Vivacity, Domestic industry
Garden Marygold	Uneasiness	Houstonia	Content
Garden Ranunculus	You are rich in attractions	Hoya	Sculpture
		Hoyabella	Contentment
Garden Sage	Esteem	Humble Plant	Despondency
Garland of Roses	Reward of virtue	Hundred-leaved Rose	Dignity of mind
Gardenia	Refinement	Hyacinth	Sport, Game, Play
Germander Speedwell	Facility	Hyacinth, Purple	Sorrowful
Geranium	Deceit	Hyacinth, White	Unobtrusive loveliness
Geranium, Dark	Melancholy	Hydrangea	A boaster
Geranium, Horseshoe-leaf	Stupidity	Hyssop	Cleanliness
Geranium, Ivy	Bridal favor		
Geranium, Lemon	Unexpected meeting	Iceland Moss	Health
Geranium, Nutmeg	Expected meeting	Ice Plant	Your looks freeze me
Geranium, Oak-leaved	True friendship	Imbricata	Uprightness, Sentiments of honor
Geranium, Pencilled	Ingenuity		
Geranium, Rose-scntd.	Preference	Imperial Montague	Power
Geranium, Scarlet	Comforting	Indian Cress	Warlike trophy
Geranium, Silver-leavd.	Recall	Indian Jasmine (Ipomoea)	Attachment
Geranium, Wild	Steadfast piety		
Gillyflower	Bonds of affection	Indian Pink (Double)	Always lovely
Gladioli	Ready armed	Indian Plum	Privation
Glory Flower	Glorious beauty	Iris	Message
Goat's Rue	Reason	Iris, German	Flame
Golden Rod	Precaution		

174

Ivy	Friendship, Fidelity, Marriage	Linden or Lime Trees	Conjugal love
Ivy, Sprig of, with Tendrils	Assiduous to please	Lint	I feel my obligations
		Live Oak	Liberty
		Liverwort	Confidence
Jacob's Ladder	Come down	Liquorice, Wild	I declare against you
Japan Rose	Beauty is your only attraction	Lobelia	Malevolence
		Locust Tree	Elegance
Japanese Lilies	You cannot deceive me	Locust Tree (green)	Affection beyond the grave
Jasmine	Amiability		
Jasmine, Cape	Transport of joy	London Pride	Frivolity
Jasmine, Carolina	Separation	Lote Tree	Concord
Jasmine, Indian	I attach myself to you	Lotus	Eloquence
Jasmine, Spanish	Sensuality	Lotus Flower	Estranged love
Jasmine, Yellow	Grace and elegance	Lotus Leaf	Recantation
Jonquil	I desire a return of affection	Love in a Mist	Perplexity
		Love lies Bleedir	Hopeless, not heartless
Judas Tree	Unbelief, Betrayal	Lucern	Life
Julienne, White	Despair not: God is everywhere	Lupin	Voraciousness
Juniper	Succor, Protection	Madder	Calumny
Justicia	The perfection of female loveliness	Magnolia	Love of nature, Magnificence
		Magnolia, Swamp	Perseverance
		Mallow	Mildness
Kennedia	Mental beauty	Mallow, Marsh	Beneficence
Kingcups	Desire of riches	Mallow, Syrian	Consumed by love
		Mallow, Venetian	Delicate beauty
		Malon Creeana	Will you share my fortunes?
Laburnum	Forsaken, Pensive beauty		
Lady's Slipper	Capricious beauty, Win me and wear me	Manchineal Tree	Falsehood
		Mandrake	Horror
Lagerstraemia, Indian	Eloquence	Maple	Reserve
Lantana	Rigor	Marianthus	Hope for better days
Lapageria Rosea	There is no unalloyed good	Marygold	Grief
		Marygold, African	Vulgar minds
Larch	Audacity, Boldness	Marygold, French	Jealousy
Larkspur	Lightness, Levity	Marygold, Prophetic	Prediction
Larkspur, Pink	Fickleness	Marygold and Cypress	Despair
Larkspur, Purple	Hughtiness	Marjoram	Blushes
Laurel	Glory	Marvel of Peru	Timidity
Laurel, Common, in flower	Perfidy	Meadow Lychnis	Wit
		Meadow Saffron	My best days are past
Laurel, Ground	Perseverance	Meadowsweet	Uselessness
Laurel, Mountain	Ambition	Mercury	Goodness
Laurel-leavd. Magnolia	Dignity	Mesembryanthemum	Idleness
Laurestina	A token	Mezereon	Desire to please
Lavender	Distrust	Michaelmas Daisy	After-thought
Leaves (dead)	Melancholy	Mignonette	Your qualities surpss your charms
Lemon	Zest		
Lemon Blossoms	Fidelity in love	Milfoil	War
Leschenaultia Splendens	You are charming	Milkvetch	Your presence softens my pains
Lettuce	Cold-heartedness	Milkwort	Hermitage
Lichen	Dejection, Solitude	Mimosa (Sensitive Plt.)	Sensitiveness
Lilac, Field	Humility	Mint	Virtue
Lilac, Purple	First emotions of love	Mistletoe	I surmount difficulties
Lilac, White	Youthful innocence	Mitraria Coccinea	Indolence, Dulness
Lily, Day	Coquetry	Mock Orange	Counterfeit
Lily, Imperial	Majesty	Monarda Amplexicaulis	Your whims are quite unbearable
Lily, White	Purity, Sweetness		
Lily, Yellow	Falsehood, Gaiety	Monkshood	A deadly foe is near
Lily of the Valley	Return of happiness, Unconscious sweetness	Monkshood (Helmet Flower)	Chivalry, Knight-errantry
		Moonwort	Forgetfulness
		Morning Glory	Affectation

175

Moschatel	Weakness
Moss	Maternal love
Mosses	Ennui
Mossy Saxifrage	Affection
Motherwort	Concealed love
Mountain Ash	Prudence
Mourning Bride	Unfortunate attachment, I have lost all
Mouse-eared Chickweed	Ingenuous simplicity
Mouse-eared Scorpion Grass	Forget-me-not
Moving Plant	Agitation
Mudwort	Happiness, Tranquility
Mulberry Tree, Black	I shall not survive you
Mulberry Tree, White	Wisdom
Mushroom	Suspicion, or, I can't entirely trust you
Musk Plant	Weakness
Mustard Seed	Indifference
Myrobalan	Privation
Myrrh	Glasness
Myrtle	Love

Narcissus	Egotism
Nasturtium	Patriotism
Nemophila	Success everywhere
Nettle, Common Stinging	You are spiteful
Nettle, Burning	Spender
Nettle Tree	Conceit
Night-blooming Cereus	Transient beauty
Night Convolvulus	Night
Nightshade	Falsehood
Oak Leaves	Bravery
Oak Tree	Hospitality
Oak, White	Independence
Oats	The witching soul of music
Oleander	Beware
Olive	Peace
Orange Blossoms	Your purity equals your loveliness
Orange Flowers	Chastity, Bridal festivities
Orange Tree	Generosity
Orchis	A belle
Osier	Frankness
Osmunda	Dreams
Ox Eye	Patience

Palm	Victory
Pansy	Thoughts
Parsley	Festivity
Pasque Flower	You have no claims
Passion Flower	Religious superstition, when the flower is reversed, or Faith if erect
Patience Dock	Patience
Pea, Everlasting	An appointed meeting, Lasting pleasure
Pea, Sweet	Departure
Peach	Your qualities, like your charms, are unequalled
Peach Blossom	I am your captive
Pear	Affection
Pear-tree	Comfort
Penstemon Azureum	High-bred
Pennyroyal	Flee away
Peony	Shame, Bashfulness
Peppermint	Warmth of feeling
Periwinkle, Blue	Early friendship
Periwinkle, White	Pleasures of memory
Persicaria	Restoration
Persimon	Bury me amid Nature's beauties
Peruvian Heliotrope	Devotion
Petunia	Your presence soothes me
Pheasant's Eye	Remembrance
Phlox	Unanimity
Pigeon Berry	Indifference
Pimpernel	Change, Assignation
Pine	Pity
Pine-apple	You are perfect
Pine, Pitch	Philosophy
Pine, Spruce	Hope in adversity
Pink	Boldness
Pink, Carnation	Woman's love
Pink, Indian, Double	Always lovely
Pink, Indian, Single	
Pink, Mountain	Aspiring
Pink, Red, Double	Pure and ardent love
Pink, Single	Pure love
Pink, Variegated	Refusal

Pink, White	Ingeniousness, Talent
Plantain	White man's footsteps
Plane Tree	Genius
Plum, Indian	Privation
Plum Tree	Fidelity
Plum, Wild	Independence
Plumbago Larpenta	Holy wishes
Polyanthus	Pride of riches
Polyanthus, Crimson	The heart's mystery
Polyanthus, Lilac	Confidence
Pomegranate	Foolishness
Pomegranate Flower	Mature elegance
Poor Robin	Compensation, or an Equivalent
Poplar, Black	Courage

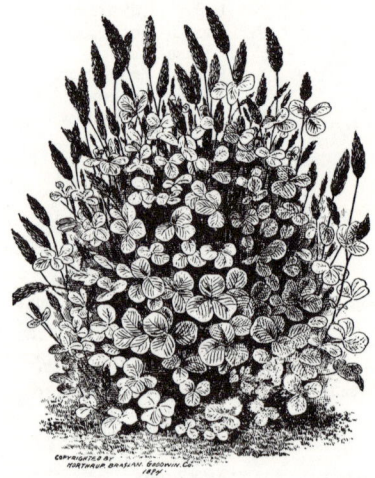

Poplar, White	Time
Poppy, Red	Consolation
Poppy, Scarlet	Fantastic extravagance
Poppy, White	Sleep, My bane
Potato	Benevolence
Potentilla	I claim, at least, your esteem
Prickly Pear	Satire
Pride of China	Dissension
Primrose	Early youth and sadness
Primrose, Evening	Inconstancy
Primrose, Red	Unpatronized merit
Privet	Prohibition
Purple Clover	Provident
Pyrus Japonica	Fairies' fire
Quaking Grass	Agitation
Quamoclit	Busybody
Queen's Rocket	You are the queen of coquettes, Fashion
Quince	Temptation

Ragged-Robin	Wit	Rose, York & Lancaster	War	Speedwell, Germander	Facility
Ranunculus	You are radiant with charms	Rose, Fll.-blwn., placed over two Buds	Secrecy	Speedwell, Spiked	Semblance
				Spider Ophrys	Adroitness
Ranunculus, Garden	You are rich in attractions	Rose, White and Red together	Unity	Spiderwort	Esteem, not love
				Spiked Willow Herb	Pretension
Ranunculus, Wild	Ingratitude	Roses, Crown of	Reward of virtue	Spindle Tree	Your charms are engraven on my heart.
Raspberry	Remorse	Rosebud, Red	Pure and lovely		
Ray Grass	Vice	Rosebud, White	Girlhood	Star of Bethlehem	Purity
Red Catchfly	Youthful love	Rosebud, Moss	Confession of love	Starwort	After-thought
Reed	Complaisance, Music	Rose Leaf	You may hope	Starwort, American	Cheerfulness in old age
Reed, Split	Indiscretion	Rosemary	Remembrance	Stephanotis	Will you accompany me to the East?
Rhododendron (Rose-bay)	Danger, Beware	Rudbeckia	Justice		
Rhubarb	Advice	Rue	Disdain	St. John's Wort	Superstition
Rocket	Rivalry	Rush	Docility	Stock	Lasting beauty
Rose	Love	Rye Grass	Changeable disposition	Stock, Ten-week	Promptness
Rose, Austrian	Thou art all that is lovely			Stonecrop	Tranquillity
				Straw (broken)	Rupture of a contract
Rose, Bridal	Happy love			Straw (whole)	Union
Rose, Burgundy	Unconscious beauty			Strawberry Blossoms	Foresight
Rose, Cabbage	Ambassador of love			Strawberry Tree	Esteem, not love
Rose, Campion	Only deserve my love			Sultan, Lilac	I forgive you
Rose, Carolina	Love is dangerous			Sultan, White	Sweetness
Rose, China	Beauty always new			Sultan, Yellow	Contempt
Rose, Christmas	Tranquillize my anxiety			Sumach, Venice	Splendor
				Sunflower, Dwarf	Adoration
Rose, Daily	Thy smile I aspire to			Sunflower, Tall	Haughtiness, False riches
Rose, Damask	Brilliant complexion				
Rose, Deep Red	Bashful shame			Swallow-wort	Cure for heartache
Rose, Dog	Pleasure and pain			Sweet Basil	Good wishes
Rose, Guelder	Winter, Age			Sweetbriar, American	Simplicity
Rose, Hundred-leaved	Pride			Sweetbriar, European	I would to heal
Rose, Japan	Beauty is your only attraction			Sweetbriar, Yellow	Decrease of love
				Sweet Pea	Delicate pleasures
				Sweet Sultan	Felicity
				Sweet William	Gallantry, Dexterity
				Sycamore	Curiosity
				Syringa	Memory, Fraternal sympathy
		Saffron	Beware of excess		
		Saffron Crocus	Mirth	Syringa, Carolina	Disappointment
		Saffron, Meadow	My happiest days are past		

		Sage	Domestic virtue	
		Sage, Garden	Esteem	
		Sainfoin	Agitation	
		Saint John's Wort	Animosity	
		Salvia, Blue	Wisdom	
		Salvia, Red	Energy	
		Saxifrage, Mossy	Affection	
		Scabious	Unfortunate love	
		Scabious, Sweet	Widowhood	
		Scarlet Lychnis	Sunbeaming eyes	
		Schinus	Religious enthusiasm	
		Scotch Fir	Elevation	
		Sensitive Plant	Sensibility	
		Senvy	Indifference	
		Shamrock	Light-heartedness	
		Shepherd's Purse	I offer you my all	
Rose, Maiden-blush	If you love me you will find it out	Sipnocampylos	Resolved to be noticed	
Rose, Montiflora	Grace	Snakesfoot	Horror	
Rose, Mundi	Variety	Snapdragon	Presumption, also 'No'	
Rose, Musk	Capricious beauty	Snowball	Bound	
Rose, Musk, Cluster	Charming	Snowdrop	Hope	
Rose, Single	Simplicity	Sorrel	Affection	
Rose, Thornless	Early Attachment	Sorrel, Wild	Wit ill-timed	
Rose, Unique	Call me not beautiful	Sorrel, Wood	Joy	
Rose, White	I am worthy of you	Southernwood	Jest, Bantering	
Rose, White (withered)	Transient impressions	Spanish Jasmine	Sensuality	
Rose, Yellow	Decrease of love. Jealousy	Spearmint	Warmth of sentiment	
		Speedwell	Female fidelity	

177

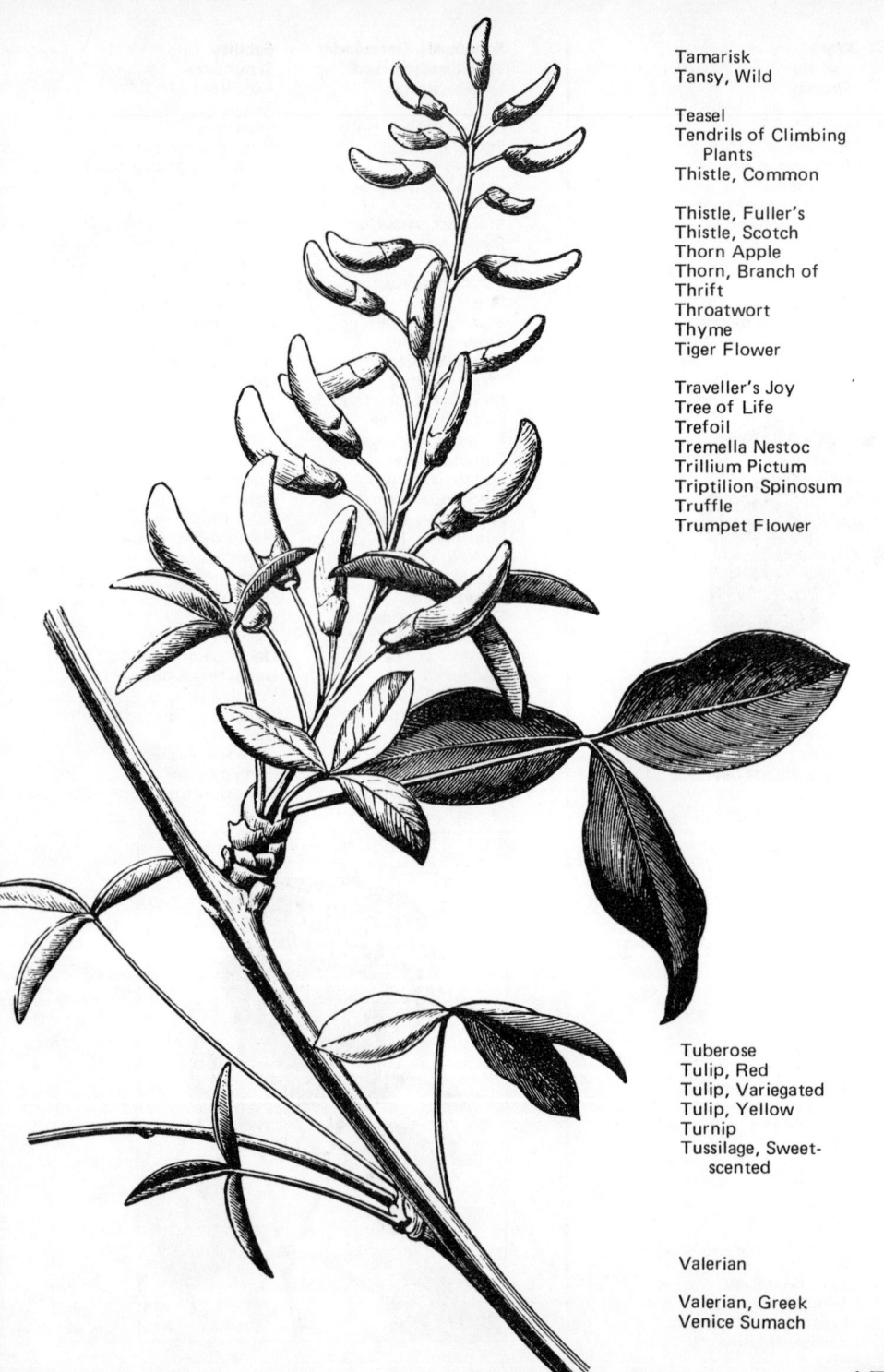

Tamarisk	Crime
Tansy, Wild	I declare war against you
Teasel	Misanthropy
Tendrils of Climbing Plants	Ties
Thistle, Common	Austerity, Independence
Thistle, Fuller's	Misanthropy
Thistle, Scotch	Retaliation
Thorn Apple	Deceitful charms
Thorn, Branch of	Severity
Thrift	Sympathy
Throatwort	Neglected beauty
Thyme	Activity or Courage
Tiger Flower	For once may pride befriend me
Traveller's Joy	Safety
Tree of Life	Old age
Trefoil	Revenge
Tremella Nestoc	Resistance
Trillium Pictum	Modest beauty
Triptilion Spinosum	Be prudent
Truffle	Surprise
Trumpet Flower	Fame
Tuberose	Dangerous pleasures
Tulip, Red	Declaration of love
Tulip, Variegated	Beautiful eyes
Tulip, Yellow	Hopeless love
Turnip	Charity
Tussilage, Sweet-scented	Justice shall be done you
Valerian	An accommodating disposition
Valerian, Greek	Rupture
Venice Sumach	Intellectual excellence, Splendor
Venus's Car	Fly with me
Venus's Looking-glass	Flattery
Venus's Trap	Deceit
Verbena, Pink	Family union
Verbena, Scarlet	Unite against evil, or Church unity
Verbena, White	Pray for me
Vernal Grass	Poor, but happy
Veronica	Fidelity
Veronica Speciosa	Keep this for my sake
Vervain	Enchantment
Vine	Intoxication
Violet, Blue	Faithfulness
Violet, Dame	Watchfulness
Violet, Sweet	Modesty
Violet, Yellow	Rural happiness
Virginia Creeper	I cling to you both in sunshine and shade
Virgin's Bower	Filial love
Viscaria Oculata	Will you dance with me?
Volkamenia	May you be happy
Walnut	Intellect, Stratagem
Wallflower	Fidelity in adversity
Watcher by the Wayside	Never despair
Water-Lily	Purity of heart
Water-Melon	Bulkiness
Wax Plant	Susceptibility
Wheat Stalk	Riches
Whin	Anger
White Flytrap	Deceit
White Jasmine	Amiableness
White Lily	Purity and modesty
White Mullein	Good-nature
White Oak	Independence
White Pink	Talent
White Poplar	Time
White Rose (dried)	Death preferable to loss of innocence
Whortleberry	Treason
Willow, Creeping	Love forsaken
Willow, Water	Freedom
Willow, Weeping	Mourning
Willow Herb	Pretension
Willow, French	Bravery and humanity
Winter Cherry	Deception
Wisteria	Welcome, fair stranger
Witch Hazel	A spell
Woodbine	Fraternal love
Wood Sorrel	Joy, Maternal tenderness
Wormwood	Absence
Xanthium	Rudeness, Pertinacity
Xeranthemum	Cheerfulness under adversity
Yew	Sorrow
Zephyr Flower	Expectation
Zinnia	Thoughts of absent friends

178

Absence	Wormwood	Beauty, Modest	Trillium Pictum	Conjugal love	Lime or Linden Tree
Abuse not	Crocus	Beauty, Neglected	Throatwort	Consolation	Red Poppy
Acknowledgment	Canterbury Bell	Beauty, Pensive	Laburnum	Constancy	Bluebell
Activity or Courage	Thyme	Beauty, Rustic	French Honeysuckle		
A deadly foe is near	Monkshood	Beauty, Unconscious	Burgundy Rose		

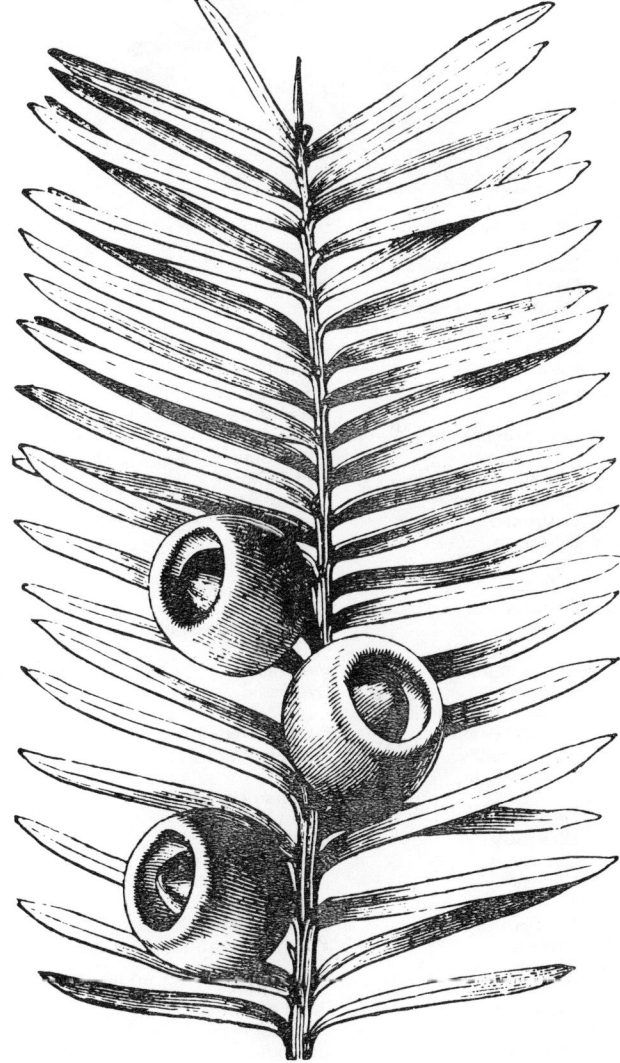

Absence — Wormwood
Abuse not — Crocus
Acknowledgment — Canterbury Bell
Activity or Courage — Thyme
A deadly foe is near — Monkshood
Admiration — Amethyst
Adoration — Dwarf Sunflower
Adroitness — Spider Ophrys
Adulation — Cacalia
Advice — Rhubarb
Affection — Mossy Saxifrage
Affection — Pear
Affection — Sorrel
Affection beyond the grave — Green Locust
Affection, Maternal — Cinquefoil
Affectation — Cockscomb Amaranth
Affectation — Morning Glory
Affliction — Black Poplar
After-thought — Michaelmas Daisy
After-thought — Starwort
After-thought — China Aster
Agreement — Straw
Age — Guelder Rose
Agitation — Moving Plant
Agitation — Sainfoin
Alas! for my poor heart — Deep Red Carnation
Always cheerful — Coreopsis
Always delightful — Cineraria
Always lovely — Indian Pink (double)
Ambassador of Love — Cabbage Rose
Amiability — Jasmine
Anger — Whin
Anger — Furze
Animosity — St. John's Wort
Anticipation — Gooseberry
Anxious and trembling — Red Columbine
Ardor, Zeal — Cuckoo Plant
Argument — Fig
Arts — Acanthus
Artifice — Clematis
Assiduous to please — Sprig of ivy, with tendrils
Attachment — Indian Jasmine
Audacity — Larch
Avarice — Scarlet Auricula
Aversion — Chain or Indian Pink

Bantering — Southernwood
Baseness — Dodder of Thyme
Bashfulness — Peony
Bashful shame — Deep Red Rose
Be prudent — Triptilion Spinosum
Be warned in time — Echites Atro-purpurea
Beautiful eyes — Variegated Tulip
Beauty — Parti-colored Daisy
Beauty always new — China Rose
Beauty, Capricious — Lady's Slipper
Beauty, Capricious — Musk Rose
Beauty, Delicate — Flower of an hour
Beauty, Delicate — Hibiscus
Beauty, Divine — American Cowslip
Beauty, Glorious — Glory Flower
Beauty, Lasting — Stock
Beauty, Magnificent — Calla Aethiopica
Beauty, Mental — Clematis

Beauty, Modest — Trillium Pictum
Beauty, Neglected — Throatwort
Beauty, Pensive — Laburnum
Beauty, Rustic — French Honeysuckle
Beauty, Unconscious — Burgundy Rose
Beauty is your only attraction — Japan Rose
Belle — Orchis
Be mine — Four-leaved Clover
Beneficence — Marshmallow
Benevolence — Potato
Betrayed — White Catchfly
Beware — Oleander
Beware — Rosebay
Beware of a false friend — Francisca Latifolia
Bitterness — Aloe
Blackness — Ebony Tree
Bluntness — Borage
Blushes — Marjoram
Boaster — Hydrangea
Boldness — Pink
Bonds — Convolvulus
Bonds of affection — Gillyflower
Bravery — Oak Leaves
Bravery and humanity — French Willow
Bridal favor — Ivy Geranium
Brilliant complexion — Damask Rose
Bulk — Water-Melon
Bulk — Gourd
Busybody — Quamoclit
Bury me amid Nature's beauties — Persimon

Call me not beautiful — Rose, Unique
Calm repose — Buckbean
Calumny — Hellebore
Calumny — Madder
Change — Pimpernel
Changeable disposition — Rye Grass
Charity — Turnip
Charming — Cluster of Musk Roses
Charms, Deceitful — Thorn Apple
Cheerfulness — Saffron Crocus
Cheerfulness in old age — American Starwort
Cheerfulness under adversity — Chinese Chrysanthemum
Chivalry — Monkshood
Cleanliness — Hyssop
Cold-heartedness — Lettuce
Coldness — Agnus Castus
Color of my life — Coral Honeysuckle
Come down — Jacob's Ladder
Comfort — Pear Tree
Comforting — Scarlet Geranium
Compassion — Allspice
Concealed love — Motherwort
Concert — Nettle Tree
Concord — Lote Tree
Confession of love — Moss Rosebud
Confidence — Hepatica
Confidence — Lilac Polyanthus
Confidence — Liverwort
Confidence in Heaven — Flowering Reed

Conjugal love — Lime or Linden Tree
Consolation — Red Poppy
Constancy — Bluebell

Consumed by love — Syrian Mallow
Contentment — Hoyabella
Could you bear poverty? — Browallia Jamisonii
Counterfeit — Mock Orange
Courage — Black Poplar
Crime — Tamarisk
Cure — Balm of Gilead
Cure for heartache — Swallow-wort
Curiosity — Sycamore

179

Danger	Rhododendron, Rosebay
Dangerous pleasures	Tuberose
Death	Cypress
Death preferable to loss of innocence	White Rose (dried)
Deceit	Apocynum
Deceit	White Flytrap
Deceit	Dogsbane
Deceit	Geranium
Deceitful charms	Apple, Thorn
Deception	White Cherry Tree
Declaration of love	Red Tulip
Decrease of love	Yellow Rose
Deformed	Begonia
Dejection	Lichen
Delay	Eupatorium
Delicacy	Bluebottle, Centaury
Delicacy	Cornflower
Depart	Dandelion Seeds in the ball
Desire to please	Mezereon
Despair	Cypress
Despair not, God is everywhere	White Julienne
Despondency	Humble Plant
Devotion, or, I turn to thee	Peruvian Heliotrope
Dexterity	Sweet William
Difficulty	Blackthorn
Dignity	Cloves
Dignity	Laurel-leaved Magnolia
Disappointment	Syringa, Carolina
Disdain	Yellow Carnation
Disdain	Rue
Disgust	Frog Ophrys
Dissension	Pride of China
Distinction	Cardinal Flower
Distrust	Lavender
Divine beauty	American Cowslip
Docility	Rush
Domestic industry	Flax
Domestic virtue	Sage
Do not despise my poverty	Shepherd's Purse
Do not refuse me	Escholzia, or Carrot Flower
Doubt	Apricot Blossom
Durability	Dogwood
Duration	Cornel Tree

Early attachment	Thornless Rose
Early friendship	Blue Periwinkle
Early youth	Primrose
Elegance	Locust Tree
Elegance and grace	Yellow Jasmine
Elevation	Scotch Fir
Eloquence	Lagerstraemia, Ind.
Enchantment	Holly Herb
Enchantment	Vervain
Energy	Red Salvia
Energy in adversity	Camomile
Envy	Bramble
Error	Bee Orchis
Error	Fly Orchis
Esteem	Garden Sage
Esteem, not love	Spiderwort
Esteem, not love	Strawberry Tree
Estranged love	Lotus Flower
Excellence	Camellia Japonica
Expectation	Anemone
Expectation	Zephyr Flower
Expected meeting	Nutmeg Geranium
Extent	Gourd
Extinguished hopes	Major Convolvulus

Facility	Germander Speedwell
Fairies' fire	Pyrus Japonica
Faithfulness	Blue Violet
Faithfulness	Heliotrope
Falsehood	Bugloss, Deadly Nightshade
Falsehood	Yellow Lily
Falsehood	Manchineal Tree
False riches	Tall Sunflower
Fame	Tulip
Fame speaks him great and good	Apple Blossom
Family union	Pink Verbena
Fantastic extravagance	Scarlet Poppy
Farewell	Michaelmas Daisy
Fascination	Fern
Fascination	Honesty
Fashion	Queen's Rocket
Fecundity	Hollyhock
Felicity	Sweet Sultan
Female fidelity	Speedwell
Festivity	Parsley
Fickleness	Abatina
Fickleness	Pink Larkspur
Filial love	Virgin's-bower
Fidelity	Veronica, Ivy
Fidelity	Plum Tree
Fidelity in adversity	Wallflower
Fidelity in love	Lemon Blossoms
Fire	Fleur-de-Luce
First emotions of love	Purple Lilac
Flame	Fleur-de-lis, Iris
Flattery	Venus's Looking-glass
Flee away	Pennyroyal
Fly with me	Venus's Car
Folly	Columbine
Foppery	Cockscomb, Amaranth
Foolishness	Pomegranate
Foresight	Holly

Forgetfulness	Moonwort
Forget me not	Forget-me-not
For once may pride befriend me	Tiger Flower
Forsaken	Garden Anemone
Forsaken	Laburnum
Fortitude	Dipteracanthus Spectabilis
Fragrance	Camphire
Frankness	Osier
Fraternal love	Woodbine
Fraternal sympathy	Syringa
Freedom	Water Willow
Freshness	Damask Rose
Friendship	Acacia, Ivy
Friendship, early	Blue Periwinkle
Friendship, true	Oak-leaved Geranium
Friendship, unchanging	Arbor Vitae
Frivolity	London Pride
Frugality	Chicory, Endive

Gaiety	Butterfly Orchis
Gaiety	Yellow Lily
Gallantry	Sweet William
Generosity	Orange Tree
Generous and devoted affection	
Genius	French Honeysuckle
Gentility	Plane Tree
Girlhood	Corn Cockle
Give me your good wishes	White Rosebud
Gladness	Sweet Basil
Glory	Myrrh
Glory, Immortality	Laurel
Glorious beauty	Daphne
Goodness	Glory Flower
Goodness	Bonus Henricus
Good education	Mercury
Good wishes	Cherry Tree
Good-nature	Sweet Basil
Gossip	White Mullein
Grace	Coboea
Grace and elegance	Multiflora Rose
Grandeur	Yellow Jasmine
Gratitude	Ash Tree
Grief	Small White Bellflower
Grief	Harebell
	Marygold
Happy love	Bridal Rose
Hatred	Basil
Haughtiness	Purple Larkspur
Haughtiness	Tall Sunflower
Health	Iceland Moss
Hermitage	Milkwort
Hidden worth	Coriander
High-bred	Pentstemon Azureum
Holy wishes	Plumbago Larpenta
Honesty	Honesty
Hope	Flowering Almond
Hope	Hawthorn
Hope	Snowdrop
Hope in adversity	Spruce Pine
Hopeless love	Yellow Tulip
Hopeless, not heartless	Love lies bleeding
Horror	Mandrake
Horror	Dragonswort
Horror	Snakesfoot

Hospitality	Oak Tree
Humility	Broom
Humility	Bindweed, Small
Humility	Field Lilac

I am too happy	Cape Jasmine
I am your captive	Peach Blossom
I am worthy of you	White Rose
I change but in death	Bay Leaf
I claim at least your esteem	Potentilla
I dare not	Veronica Speciosa
I declare against you	Belvidere
I declare against you	Liquorice
I declare war against you	Wild Tansy
I die if neglected	Laurestina
I desire a return of affection	Jonquil
I feel my obligations	Lint
I feel your kindness	Flax
I have lost all	Mourning Bride
I live for thee	Cedar Leaf
I love	Red Chrysanthemum
I offer you my all	Shepherd's Purse
I offer you my fortune, or, I offer you pecuniary aid	Calceolaria
I share your sentiments	Double China Aster
I share your sentiments	Garden Daisy
I shall die tomorrow	Gum Cistus
I shall not survive you	Black Mulberry
I surmount difficulties	Mistletoe
I watch over you	Mountain Ash
I weep for you	Purple Verbena
I will think of it	Single China Aster
I will think of it	Wild Daisy
I would to heal	Eglantine, Sweetbriar
If you love me, you will find it out	Maiden Blush Rose
Idleness	Mesembryanthemum
Ill nature	Crab Blossom
Ill natured beauty	Citron
Imagination	Lupine
Immortality	Amaranth, Globe
Impatience	Yellow Balsam

Impatient of absence	Corchorus
Impatient resolves	Red Balsam
Imperfection	Henbane
Importunity	Burdock
Inconstancy	Evening Primrose
Incorruptible	Cedar of Lebanon
Independence	Common Thistle
Independence	Wild Plum Tree
Independence	White Oak
Indifference	Candytuft, Everflowering
Indifference	Mustard Seed
Indifference	Pigeon Berry
Indifference	Senvy
Indiscretion	Split Reed
Indolence	Mittraria Coccinea
Industry	Red Clover
Industry, Domestic	Flax
Ingeniousness	White Pink
Ingenuity	Pencilled Geranium
Ingenuous simplicity	Mouse-eared Chickweed
Ingratitude	Crowfoot
Innocence	Daisy
Insincerity	Foxglove
Insinuation	Great Bindweed
Inspiration	Angelica
Instability	Dahlia
Intellect	Walnut
Intoxication	Vine
Irony	Sardony

Jealousy	French Marygold
Jealousy	Yellow Rose
Jest	Southernwood
Joy	Wood Sorrel
Joys to come	Lesser Celandine
Justice	Rudbeckia
Justice shall be done to you	Coltsfoot, or Sweet-scented Tussilage
Keep your promise	Petunia
Kindness	Scarlet Geranium
Knight-errantry	Helmet Flower (Monkshood)
Lamentation	Aspen Tree
Lasting beauty	Stock
Lasting pleasures	Everlasting Pea
Let me go	Butterfly Weed
Levity	Larkspur
Liberty	Live Oak
Life	Lucern
Light-heartedness	Shamrock
Lightness	Larkspur
Live for me	Arbor Vitae
Love	Myrtle
Love	Rose
Love, forsaken	Creeping Willow
Love, returned	Ambrosia
Love is dangerous	Carolina Rose
Love for all seasons	Furze
Lustre	Aconite-leaved Crowfoot, or Fair Maid of France
Luxury	Chestnut Tree
Magnificence	Magnolia
Magnificent beauty	Calla Aethiopica
Majesty	Crown Imperial
Make haste	Dianthus
Malevolence	Lobelia

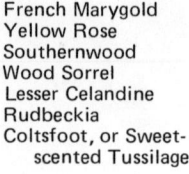
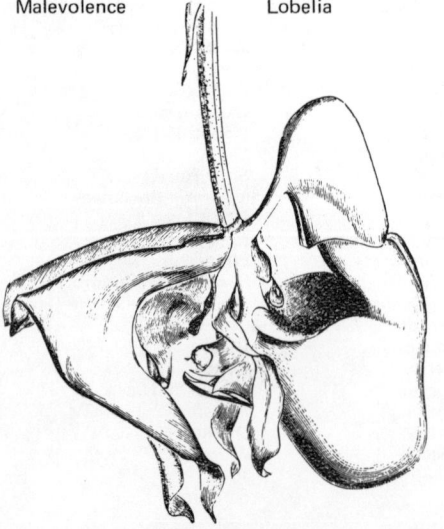

Marriage	Ivy
Maternal affection	Cinquefoil
Maternal love	Moss
Maternal tenderness	Wood Sorrel
Matrimony	American Linden
Matronly grace	Cattleya
Mature charms	Cattleya Pineli
May you be happy	Volkamenia
Meanness	Cuscuta
Meekness	Birch
Melancholy	Autumnal Leaves
Melancholy	Dark Geranium
Melancholy	Dead Leaves
Mental beauty	Clematis
Mental beauty	Kennedia
Message	Iris
Mildness	Mallow
Mirth	Saffron Crocus
Misanthropy	Aconite (Wolfsbane)
Misanthropy	Fuller's Teazle
Modest beauty	Trillium Pictum
Modest genius	Creeping Cereus
Modesty	Violet
Modesty and purity	White Lily
Momentary happiness	Virginian Spiderwort
Mourning	Weeping Willow
Music	Bundles of Reed with their Panicles
My best days are past	Colchicum, or Meadow Saffron
My regrets follow you to the grave	Asphodel
Neatness	Broom
Neglected beauty	Throatwort
Never-ceasing remembrance	Everlasting
Never despair	Watcher by the Wayside
No	Snapdragon
Old age	Tree of life
Only deserve my love	Campion Rose
Painful recollections	Flos Adonis
Painting	Auricula
Painting the lily	Daphne Odora
Passion	White Dittany
Paternal error	Cardamine
Patience	Dock, Ox-eye
Patriotism	American Elm
Patriotism	Nasturtium
Peace	Olive
Perfected loveliness	Camellia Japonica, White
Perfidy	Common Laurel, in flower
Pensive beauty	Laburnum
Perplexity	Love in a Mist
Persecution	Chequered Fritillary
Perseverance	Swamp Magnolia

Persuasion	Althea Frutex	Riches	Corn	Sadness	Dead Leaves
Persuasion	Syrian Mallow	Riches	Buttercups	Safety	Traveller's Joy
Pertinacity	Clotbur	Rigour	Lantana	Satire	Prickly Pear
Pity	Pine, also Andromeda	Rivalry	Rocket	Sculpture	Hoya
Pleasure and pain	Dog Rose	Rudeness	Clotbur	Secret love	Yellow Acacia
Pleasure, lasting	Everlasting Pea	Rudeness	Xanthium	Semblance	Spiked Speedwell
Pleasures of memory	White Periwinkle			Sensitiveness	Mimosa
Pomp	Dahlia			Sensuality	Spanish Jasmine
Popular favor	Cistus, or Rock Rose			Separation	Carolina Jasmine
Poverty	Evergreen Clematis			Severity	Branch of Thorns
Power	Imperial Montague			Shame	Peony
Power	Cress			Sharpness	Barberry Tree
Pray for me	White Verbena			Sickness	Anemone (Zephyr Fir)
Precaution	Golden Rod			Silent love	Evening Primrose
Prediction	Prophetic Marygold			Silliness	Fool's Parsley
Pretension	Spiked Willow Herb			Simplicity	American Sweetbriar
Pride	Hundred-leaved Rose			Sincerity	Garden Chervil
Pride	Amaryllis			Slighted love	Yel. Chrysanthemum
Privation	Indian Plum	Rural happiness	Yellow Violet	Snare	Catchfly, Dragon Plant
Privation	Myrobalan	Rustic beauty	French Honeysuckle	Solitude	Heath
Profit	Cabbage	Rustic oracle	Dandelion	Sorrow	Yew
Prohibition	Privet			Sourness of temper	Barberry
Prolific	Fig Tree			Spell	Circaea
Promptness	Ten-week Stock			Spleen	Fumitory
Prosperity	Beech Tree			Splendid beauty	Amaryllis
Protection	Bearded Crepis			Splendor	Austurtium
Prudence	Mountain Ash			Sporting	Fox-tail Grass
Pure love	Single Red Pink			Steadfast piety	Wild Geranium
Pure and ardent love	Double Red Pink			Stoicism	Box Tree
Pure and lovely	Red Rosebud			Strength	Cedar, Fennel
Purity	Star of Bethlehem			Stupidity	Horseshoe-leaf Geranium
				Submission	Grass
Quarrel	Broken Corn-straw			Submission	Harebell
Quicksightedness	Hawkweed			Success everywhere	Nemophila
				Success crown your wishes	Coronella
Ready-armed	Gladioli			Succor	Juniper
Reason	Goat's Rue			Such worth is rare	Achimenes
Recantation	Lotus Leaf			Sun-beaming eyes	Scarlet Lychnis
Recall	Silver-lvd. Geranium			Superstition	St. John's Wort
Reconciliation	Filbert			Surprise	Truffle
Reconciliation	Hazel			Susceptibility	Wax Plant
Refinement	Gardenia			Suspicion	Champignon
Refusal	Striped Carnation			Sympathy	Balm
Regard	Daffodil			Sympathy	Thrift
Regret	Purple Verbena				
Relief	Balm of Gilead				
Relieve my anxiety	Christmas Rose			Talent	White Pink
Religious superstition	Aloe			Tardiness	Flax-leaved Golden-locks
Religious superstition, or Faith	Passion Flower			Taste	Scarlet Fuchsia
Religious enthusiasm	Schinus			Tears	Helenium
Remembrance	Rosemary			Temperance	Azalea
Remorse	Bramble			Temptation	Apple
Remorse	Raspberry			Thankfulness	Agrimony
Rendezvous	Chickweed			The color of my fate	Coral Honeysuckle
Reserve	Maple			The heart's mystery	Crimson Polyanthus
Resistance	Tremella Nestoe			The perfection of female loveliness	Justicia
Resolved to be noticed	Siphocampylos			The witching soul of music	Oats
Restoration	Persicaria			The variety of your conversation delights me	Clarkia
Retaliation	Scotch Thistle			Thee only do I love	Arbutus
Return of happiness	Lily of the Valley			There is no unalloyed good	Lapagenia Rosea
Revenge	Birdsfoot Trefoil				
Reverie	Flowering Fern				
Reward of merit	Bay Wreath				
Reward of virtue	Garland of Roses				

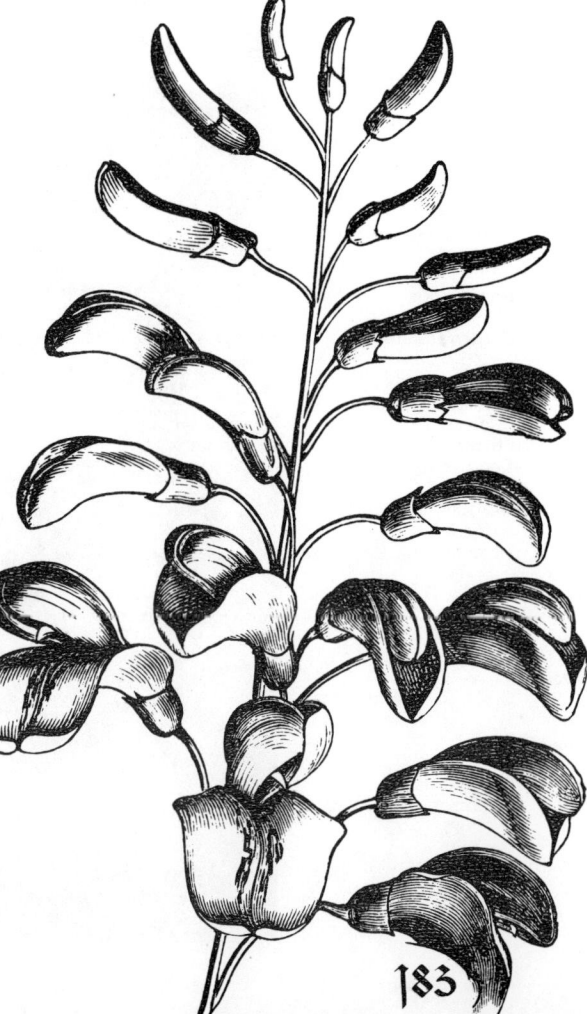

183

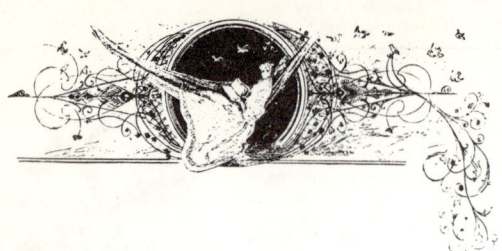

Thoughts	Pansy
Thoughts of absent friends	Zinnia
Thy frown will kill me	Currant
Thy smile I aspire to	Daily Rose
Ties	Tendrils of Climbing Plants
Timidity	Amaryllis
Timidity	Marvel of Peru
Time	White Poplar
Tranquillity	Mudwort
Tranquillity	Stonecrop
Tranquillize my anxiety	Christmas Rose
Transient beauty	Night-blooming Cereus
Transient impressions	Withered White Rose
Transport of joy	Cape Jasmine
Treachery	Bilberry
True love	Forget-me-not
True friendship	Oak-leaved Geranium
Truth	Bittersweet Nightshade
Truth	Wht. Chrysanthemum
Unanimity	Phlox
Unbelief	Judas Tree
Unceasing remembrance	American Cudweed
Unchanging friendship	Arbor Vitae
Unconscious beauty	Burgundy Rose
Unexpected meeting	Lemon Geranium
Unfortunate attachment	Mourning Bride
Unfortunate love	Scabious
Union	Whole Straw
Unity	White and Red Rose together
Unite against a common foe	Scarlet Verbena
Unpatronized merit	Red Primrose
Unrequited love	Daffodil
Uprightness	Imbricata
Uselessness	Meadowsweet
Utility	Grass
Variety	China Aster
Variety	Mundi Rose
Vice	Darnel (Ray Grass)
Victory	Palm
Virtue	Mint
Virtue, domestic	Sage
Volubility	Abecedary
Voraciousness	Lupine
Vulgar minds	African Marygold

War	York and Lancaster Rose
War	Achillea Millefolia
Warlike trophy	Indian Cress
Warmth of feeling	Peppermint
Watchfulness	Dame Violet
Weakness	Moschatel
Weakness	Musk Plant
Welcome, fair stranger	Westeria
Welcome to a stranger	American Starwort
Widowhood	Sweet Scabious
Will you accompany me to the East?	Stephanotis
Will you dance with me?	Viscaria Oculata
Win me and wear me	Lady's Slipper
Winning grace	Cowslip
Winter	Guelder Rose
Wisdom	Blue Salvia
Wit	Meadow Lychnis
Wit, ill-timed	Wild Sorrel
Witchcraft	Enchanter's Nightshade
Worth beyond beauty	Sweet Elysium
Worth sustained by judicious and tender affection	Pink Convolvulus
Worldliness, self-seeking	Clianthus
Worthy of all praise	Fennel
You are cold	Hortensia
You are my divinity	American Cowslip
You are perfect	Pineapple
You are radiant with charms	Ranunculus
You are rich in attraction	Garden Ranunculus
You are the queen of coquettes	Queen's Rocket
You are charming	Leschenaultia Splendens
You have no claims	Pasque Flower
You have many lovers	Chorozema Varium
You please all	Branch of Currants
You are too bold	Dipladenia Crassinoda
You will be my death	Hemlock
Your charms are engraven on my heart	Spindle Tree
Your looks freeze me	Ice Plant
Your presence softens my pain	Milkvetch
Your purity equals your loveliness	Orange-blossoms
Your qualities, like your charms, are unequalled	Peach
Your qualities surpass your charms	Mignonette
Your temper is too hasty	Grammanthes Chloraflora
Youthful beauty	Cowslip
Youthful innocence	White Lilac
Youthful love	Red Catchfly
Your whims are unbearable	Monarda Amplexicaulis

Zealousness	Elder
Zest	Lemon

GLOSSARY

Abortive, imperfectly or not developed; hence sterile.
Abscission, the natural cutting off of members by means of a layer of separation.
Absciss-layer, a layer of separation. See above
Acaulescent, stemless, or apparently so.
Accrescent, applied to the parts connected with the flower, as the calyx, etc., which increase in size after flowering.
Accumbent, cotyledons with margins folded against the hypocotyl.
Acerosae, Alex. Braun's term for the coniferae.
Achene, a small, dry, nonsplitting, one-celled, one-seeded fruit (as in Baccharis, Cowania, and Eriogonum).
Achlamydeous, used of flowering plants which have no calyx or corolla.
Acicular, bristle- or needle-shaped.
Acrobrya, Endlicher's term for plants growing at the apex only.
Acrocarpous, said of mosses which produce their fruit (sporogonia) at the tips of their shoots.
Actinomorphic, applied to flowers which may be divided vertically into similar halves through two or more planes.
Aculei, slender, rigid prickles, growing from the bark, as in the rose.
Acuminate, gradually tapering to the apex.
Acute, sharp pointed.
Adhesion, the union of parts normally separate.
Adnate, congenitally united or grown together.
Adventitious buds, buds produced out of their regular order.
Adventive, not indigenous, but apparently becoming naturalized.
Aecidium, in uredineae a cup-like collection of spores which are budded off from the base of the cup.
Aestivation, the folding of the parts of a flower in the bud.
After-ripening, complex biochemical or physical changes occurring in seeds, bulbs, tubers, and fruits after harvesting, when ripe in the ordinary sense, and often necessary for subsequent germination.
Aggregate fruit, a fruit formed by the crowding together of distinct carpels; the product of a single gynoeceum when that gynoeceum is apocarpous.
Aggregation, the condition of extreme activity of the stalk-cells of the tentacles of a Drosera-leaf, resulting from mechanical or chemical stimulation.
Akinetes, in green algae, are single cells of the thallus, whose original walls thicken, and which separate from the rest of the thallus; they correspond to the chlamydospores of fungi.
Alae, descriptive term applied to the two lateral members or wings of a papilionaceous corolla.
Albumen, any form of nutritive matter stored within the seed and about the embryo.
Albuminous, containing albumen, as in the seeds of grain, palms, etc.
Aleurone-grains, grains of nitrogenous food-material frequently stored in the reserve-tissues of seeds.
Alga, a chlorophyll-containing member of the thallophyta; one of the plants, the best known of which are called sea-weeds.
Alliaceous, Onion-like, in aspect or odor.
Alliance, a group of allied families or orders.
Alternate, not opposite; with a single leaf at each node.
Alveolate, like honeycomb; closely pitted.
Ament, a spike of imperfect flowers subtended by scarious bracts, as in the willows.
Amentaceous, having amenta or catkins; consisting of or resembling a catkin.
Amentum, a catkin. See catkin.
Amoeboid movements, constant changes of shape resembling those of the "proteus animalcule" amoeba.
Amphibious, said of plants such as can live either in the water or in the air.
Amphicarpium, an archegonium when it persists, after fertilization, as a fruit envelope.
Amphigastria, in liverworts: certain small scales or leaves on the ventral side of the oophyte generation.
Amphitropous, term applied to the partly inverted ovule.
Amplexicaul, nearly surrounding or clasping the stem: used of the leaf base in certain cases.
Amylum, starch.
Anastomose, to inosculate or run into each other; to communicate with each other like arteries and veins.

Anastomosing, connecting so as to form a well-defined network.
Anatomy, the intimate structure of plants.
Anatropous, said of that form of ovule in which, although the nucellus is straight, the micropyle is bent down to the point of attachment of the funicle, and in which the body of the ovule is united to the funicle, which latter structure is known as the raphe.
Androecium, the collective term for the stamens of a flower.
Androgonidia, the cells which in volvox give rise to spermatozoids.
Androgynous, flower clusters having staminate and pistillate flowers.
Androspores, name given to the particular zoospores which in oedogonium give rise to miniature plants, termed dwarf-males.
Anemophilous, applied to flowers whose pollen is conveyed by the agency of wind; having flowers fertilized by wind-borne pollen.
Angiosperm, a flowering plant, or member of the angiosperms, one of the two groups of seed plants and characterized by the seeds enclosed in a fruit. See gymnosperm.
Angiospermous, pertaining to the angiospermae; bearing seeds within a pericarp.
Animalcule, a vague term applied to small motile organisms in water.
Anisogametes, sexual cells, which show a differentiation into male and female.
Annulus, (1) in Agarics: the ring which often remains round the stalk (stipe), and was originally attached to the edge of the pileus; the remains of the velum partiale; (2) in the moss-capsule: the ring of cells which brings about the throwing-off of the operculum; (3) in the fern-sporangium: a conspicuous row of cells running vertically, obliquely, etc., around the sporangium, by the contraction of which dehiscence takes place.
Anther, the polliniferous part of a stamen; the sac or cavity in which the pollen is contained.
Antherid, the male organ of reproduction in pteridophyta and bryophyta.
Antheridium, a male sexual organ, usually producing motile spermatozoids.
Anthesis, period of flowering.
Anthocyanin, a purple sap-pigment frequent in foliage and flowers.
Antholysis, literally a "loosened" flower, i.e. a flower in which the various parts have become more or less foliacious, and from which inferences can be drawn as to the morphological nature of the component parts.
Anthoxanthin, the yellow pigment of flowers and fruits.
Antipodal cells, a group of three cells at the chalazal end of the embryo-sac of angiosperms.
Apetalae, dicotyledons destitute of a corolla.
Apetalous, without petals.
Apical, at the top, or referring to the top.
Apiculate, with a minute pointed tip.
Aplanospore, a non-motile asexual reproductive cell of the Green Algae.
Apocarpous, said when the carpels of a gynoeceum are separate.
Apophysis, a swelling under the base of the theca in some mosses.
Apothecium, the disc-like receptacle of an ascomycetous fungus.
Appressed, lying against another organ.
Arbor, a tree.
Arborescent, tree-like, in size or shape.
Arbuscula, a little or dwarf tree.
Archegonium, in the higher cryptogams the flask-shaped female sexual organ with neck and venter, the latter containing an egg-cell, the former canal-cells.
Archesporium, a cell or group of cells from which spore mother-cells are produced.
Archichlamydeae, a large group of dicotyledons, including the old groups polypetalae and incompletae.
Areolated, marked with little areas; divided into small areas by intersecting lines.
Areolation, the system of meshes in a network of veins.
Areole, a mesh in a network of veins.
Aril, an exterior appendage growing out from the hilum (scar or point of attachment of the seed) and covering the seed partly or wholly (as in celastrus, euonymus, and taxus).
Arillate, provided with an aril.
Aristate, tipped by an awn or bristle.
Aristulate, diminutive of aristate.

Arthrospore, a form of spore produced in the schizomycetes by the segmentation of the tubes into cells.
Arundinaceous, reed-like.
Ascending, growing obliquely upward, or upcurved.
Asciiform, like a pitcher; pitcher-shaped.
Ascidiform, a pitcher; an appendage somewhat resembling a pitcher. See pitcher.
Ascus, a form of sporangium characteristic of certain fungi. It is generally tubular and contains eight spores, the acospores.
Asexual, reproduction without union of male and female elements, such as in vegetative propagation.
Assimilation, as used here, the building of a plant-substance from the nutriment of the environment. Often restricted to the manufacture of carbo-hydrate from carbonic acid and water.
Asyngamic, used of plants which are prevented from intercrossing by the fact of their non-simultaneous periods of flowering. Nearly related species can thus inhabit the same spot without hybrids ever being formed.
Auricle, an ear-shaped appendage.
Autogamy, self-pollination, ultimately self-fertilization.
Autonomous movements, spontaneous; originating from inherent tendency.
Auxospore, the reproductive cell of a diatom.
Awn, a bristle-like appendage, especially in the glumes of grasses.
Axil, the upper angle formed by a leaf or branch with the stem.
Axile, in the axis of an organ.
Axis, essentially the stem. The root is also an axis.
Azygospore, term given to the "zygospore" when it is formed parthenogenetically with conjugation.

Baccate, berry-like.
Bacterium, one of the micro-organisms concerned in putrefaction: a term rather widely applied to any member of the schizomycetes.
Barbellate, furnished with minute barbs.
Barbs, the retrorse appendages of bristles, or the teeth on leaf-margins.
Bark, the usually hard outer investment of a perennial stem (or root) which has arisen in connection with a cork-cambium; actually it includes the products of the cork-cambium and whatsoever is external to it.
Basidium, a cell from which spores or conidia are produced by a process of abstriction.
Basifixed, attached by the base.
Bast, inner bark; a special tissue: soft-bast, the phloem—includes sieve-tubes and other non-hardened phloem-elements; hard-bast, the thickened prosenchymatous elements or bast-fibres.
Bastard, a term sometimes given to a hybrid.
Bedeguar, name given to the mossy red galls on the common wild rose.
Berry, a fleshy or pulpy, usually many-seeded fruit (as in arctostaphylos, linicera, and ribes).
Bilabiate, two-lipped.
Bizzaria, a fruit, part orange, part citron.
Blade, the flat expanded part of a leaf.
Blending, a name given to a hybrid arising by the crossing of "races".
Brachydodromous, used of leaf-veins.
Bract, a modified reduced leaf from the axil of which a flower stem arises.
Bracteate, with bracts.
Bracteolate, having bractlets.
Bractlet, a secondary bract, borne on a pedicel, or immediately beneath a flower; sometimes applied to minute bracts.
Bract-scale, the lower member of the duplex scale of the female cone of pine, fir, etc.
Break back, a term used by gardeners to convey the idea of reversion. Thus flowers break back or revert to an ancestral type.
Broadcast sowing, sowing seed by scattering as uniformly as possible over an area.
Browse, twigs and shoots, with their leaves, cropped by livestock and wild animals from shrubs, trees, and woody vines.
Bud, the as yet unexpanded rudiment of a shoot; it comprehends both axial and foliar portions.
Bulb (bulbus), a bud consisting of an abbreviated axis with fleshy scale-leaves in which food-material is stored. Usually subterranean.
Bulbil, a deciduous bud, usually formed on an aerial part of a plant. Occasionally used for a little bulb.

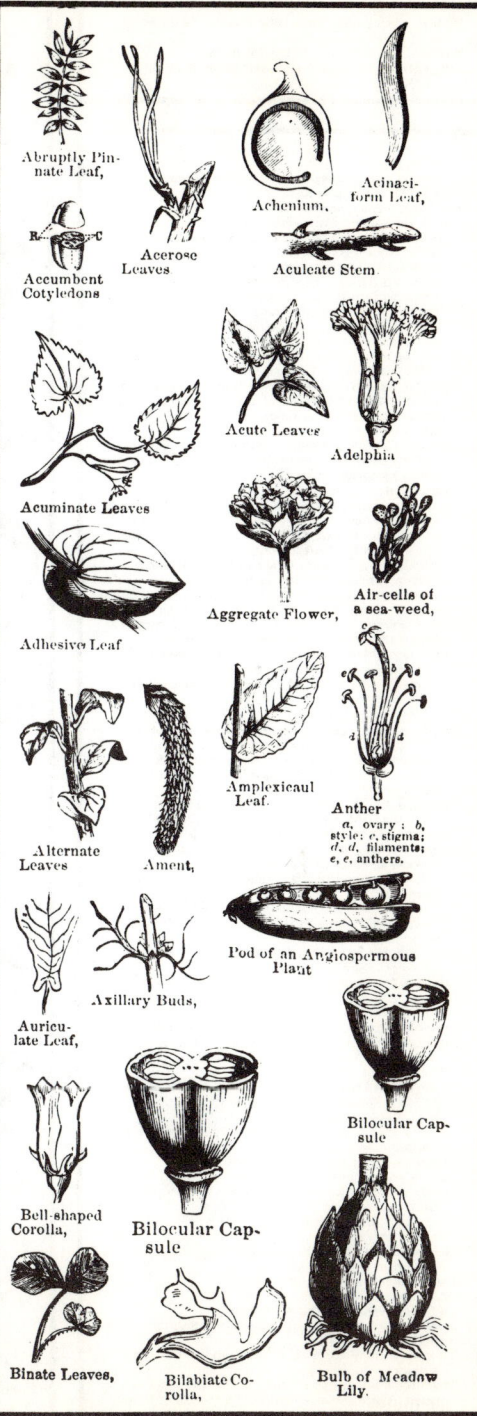

Abruptly Pinnate Leaf,
Achenium,
Acinaciform Leaf,
Accumbent Cotyledons
Acerose Leaves
Aculeate Stem
Acute Leaves
Adelphia
Acuminate Leaves
Aggregate Flower,
Air-cells of a sea-weed,
Adhesive Leaf
Amplexicaul Leaf.
Anther a, ovary: b, style; c, stigma; d, d, filaments; e, e, anthers.
Alternate Leaves
Ament,
Pod of an Angiospermous Plant
Auriculate Leaf,
Axillary Buds,
Bilocular Capsule
Bell-shaped Corolla,
Bilocular Capsule
Binate Leaves,
Bilabiate Corolla,
Bulb of Meadow Lily.

185

GLOSSARY

Bulblet, a small bulb, especially those borne on leaves, or in their axils.
Bulbous, similar to a bulb; bearing bulbs.
Bur, a prickly or spiny fruit envelope (as in castanea and fagus).

Caducous, falling away very soon after development.
Caespitose, growing in tufts.
Callosity, a small, hard protuberance.
Callus, the healing tissue which closes up the wounds of plants. The same term is given to a mucilaginous substance which arises on the sieve-plates of the sieve-tubes, closing them. The latter is of course quite a different structure, and to distinguish it from the former may be called callose.
Calyptra, the hood which is raised up on the sporogonium of a moss. It is the ruptured upper portion of the archegonium.
Calyx, the outer series of the floral envelope, or perianth; the sepals as a unit.
Cambiform cells, cells resembling cambium cells; thin-walled, tapering cells found in the phloem accompanying the sieve-tubes, compaion-cells, and bast-fibres.
Cambium, a layer of tissue formed between the wood and the bark, and consisting partly of nascent wood, partly of nascent bark.
Campanulate, bell-shaped.
Campylodromous, applied to the manner in which veins are distributed.
Campylotropous, used of an ovule or seed in which the nucellus, with its integuments, is bent so that the apex is brought near to the point of attachment.
Cancellate, reticulated, with the meshes sunken.
Canescent, with gray or hoary fine pubescence.
Canaliculate, channelled; longitudinally grooved.
Canker, a vague term applied to the disease or fungus which attacks plants and causes slow decay.
Capillitium, the thread-like fibres, often united into a reticulum, which are developed within the spores of myxomycetes and many gasteromycetes.
Capitate, arranged in a head; knob-like.
Capitulum, a head or globular cluser of sessile flowers.
Caprification, the custom of hanging branches of the wild fig in the cultivated trees so as to ensure pollination by means of the gall-insects thus introduced.
Caprificus, the uncultivated male form of the common fig.
Capsule, a dry, splitting, usually many-seeded fruit of more than one carpel (as in ceanothus and kalmia).
Carinate, keeled: with a longitudinal ridge.
Carobe di giude, turpentine gall-apple, produced on pistacia lentiscus by a pemphigus.
Carpel, a modified floral leaf, one or more of which form a pistil.
Carpium, or Carp, the oogonium modified by fertilization, which remains as an envelope around the embryo.
Carpo-asci, the more complex ascomycetous fungi—all except the exoascaceae.
Carpophylla, the carpels.
Caruncle, a localized outgrowth of the seed-coat; a sort of aril.
Caryophyllaceous, appertaining to the pink family.
Caryopsis, an indehiscent one-seeded fruit, in which the thin seed-coat adheres to the pericarp, as in all cereal grains.
Catapult-fruits, fruits in which the dispersal of the seeds or fruit-segments is due to the elastic reaction of the resilient peduncles or pedicels.
Catkin, a scaly-bracted spike of usually unisexual flowers (as in alnus and betula).
Caudate, with a slender tail-like appendage.
Caudex, a trunk or unbranched stem.
Caudex columnaris, an erect columnar stem, as in palm-trees.
Caudicle, stalk of a pollen-mass in the orchid and milk-weed families.
Caulescent, having an obvious stem rising above the ground.
Cauline, appertaining to the stem.
Caulis, the stem or stalk.
Caulis herbaceus, a herbaceous stem.
Caulis suffruticosus, a suffruticose stem; the stem of an under-shrub.
Caulome, a stem-structure, or the stem-like portion of a plant.
Cecidium, a gall or hypertrophy on a plant-member, due to the stimulating action of an insect or fungus.

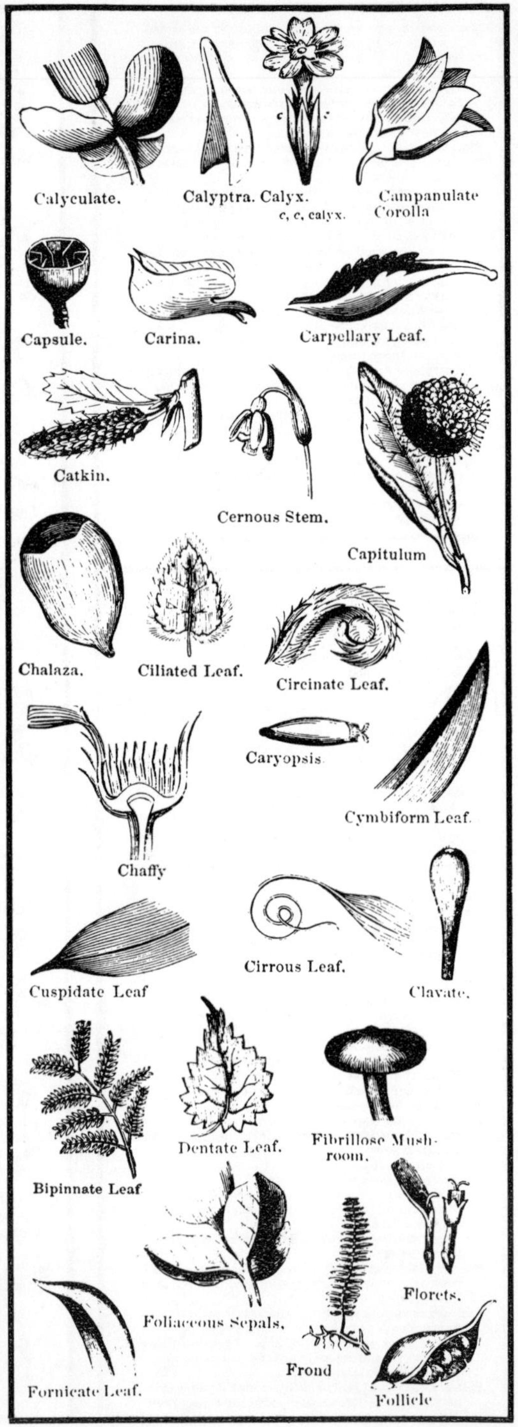

Calyculate. Calyptra. Calyx. c, c, calyx. Campanulate Corolla
Capsule. Carina. Carpellary Leaf.
Catkin. Cernous Stem. Capitulum
Chalaza. Ciliated Leaf. Circinate Leaf.
Caryopsis. Chaffy Cymbiform Leaf.
Cuspidate Leaf Cirrous Leaf. Clavate.
Bipinnate Leaf Dentate Leaf. Fibrillose Mushroom.
Foliaceous Sepals. Florets. Frond Fornicate Leaf. Follicle

Cell, the structural unit in the formation of plants; one of the individualized portions of which plants are built up.
Cell-membrane, the cell-wall.
Cell-plate, used here of aggregates of cells in one plane.
Cell-sap, the watery fluid contained in a cell.
Cellular, consisting of cells. Sometimes used of plants which are destitute of vessels.
Cellulose, a carbo-hydrate of which cell-membranes are composed; the essential constituent of cell-walls.
Centrifugal, a term applied to such inflorescences as develop from the center outwards.
Centripetal, a term applied to such inflorescences as develop from without inwards.
Cephalonion gall, a sac-like gall joined to the leaf by a narrow neck.
Ceratonion gall, a hollow, thick-walled, horn-like gall, belonging to the series of mantle-galls.
Ceresan, an organic mercury dust used for treating seeds. See uspulun.
Chaff, thin dry scales.
Chalaza, the part of an ovule where nucellus and integuments cohere; the base of the nucellus.
Chalazogamic, applied to fertilization in flowering plants via the chalaza and not by the micropyle, e.g. in the hazel.
Chartaceous, paper in texture.
Chlamydospore, the reproductive organ in some fungi.
Chloranthy, the production of green flowers; a supposed reversion of floral structures to a primitive foliar condition.
Chlorenchyma, a term sometimes given to a green, chlorophyll-containing tissue.
Chlorophyll, the ordinary green pigment of plants which is the agent in the process of carbon assimilation.
Chlorophyll-corpuscles, protoplasmic bodies distinct from, yet imbedded in, the general cell-protoplasm of the green parts of plants. The chlorophyll is restricted to these corpuscles.
Chlorophyllous, containing chlorophyll.
Chromatophore, a general term for any protoplasmic body containing a pigment. Chlorophyll-corpuscles are chromatophores.
Chromosomes, see fibrils.
Cilia, delicate protoplasmic filaments serving as organs of locomotion, as in zoospores, etc.
Ciliate, provided with marginal hairs.
Ciliolate, minutely ciliate.
Cilium, a hair.
Cincinnus, a form of cymose inflorescence, a one-sided cyme.
Cinereous, ashy; ash-colored.
Circinnate, coiled downward from the apex.
Circumscissile, transversely dehiscent, the top falling away as a lid.
Cirrhus capreolus, a term for stem-tendrils, i.e. branch-tendrils and flower-stalk tendrils.
Cirrhus costalis, a projecting or excurrent midrib, modified as a tendril.
Cirrhus foliaris, a leaf modified as a tendril.
Cirrhus peduncularis, a flower-stalk modified as a tendril.
Cirrhus petiolaris, a petiole or leaf-stalk modified as a tendril.
Cirrhus radicalis, a root modified as a tendril.
Cirrhus rameaneus, a tendril which is a modified branch.
Cirrhus stipularis, a tendril which is a metamorphosed stipule.
Cladodes, leaf-like branches. See phylloclade.
Clamp-cells, here used for the papilla-like cells by which an epiphytic root adheres to the substratum.
Class, the highest grade or division of plants in the system of Linnaeus. In our system a class is subordinate to a phylum, and the classes are subdivided into alliances.
Clavate, club-shaped.
Claw, a name given to the stalk of a petal.
Cleistogamic, -ous, a term applied to the inconspicuous flowers produced by many plants. These flowers do not open, and are self-pollinated (autogamous).
Cleft, cut about halfway to the midvein.
Clon (or clone), a group of plants propagated only by a vegetative and asexual means, all members of which have been derived by repeated propagation from a single individual.
Coalescent, two or more similar parts united or growing together.
Cochteate, like a snail shell.
Coenobel, or Coenobium, a colony of separate organisms united by a common investment, e.g. volvox.
Coherent, used of the union of similar members.

Cohort, a group of families or orders which are nearly related to one another; is used here as synonymous with alliance.
Collective fruit, a fruit in which the products of a number of separate flowers become so crowded together as to appear as though they had arisen from a single flower, as the pine-apple. Cf. aggregate fruit.
Collenchyma, a living tissue, consisting of prism-shaped cells whose angles are much thickened. It is a form of mechanical tissue.
Columella, in muscineae, the sterile tissue in the center of the sporogonium around which the spore-layer is formed.
Column, the body formed as a result of fusion of stamens with style, as in orchid flowers.
Coma, tuft of hairs at the ends of some seeds.
Commissure, the continguous surfaces of two carpels.
Conceptacle, the inclosing cavity in which the sexual organs are produced in the fucaceae.
Cone, the seed-bearing (female) or pollen-bearing (male) structure of gymnospers, including conifers, consisting of an axis with many overlapping scales (as in abies, larix, and pinus).
Conelet, a small, immature female cone.
Confluent, blended together.
Conidium, in fungi, a propagative asexual body.
Conifer, a plant producing cones; one of the coniferae.
Conjugation, the union of two gametes (or sexual cells), the resulting organism being called a zygote.
Conjugation-canal, the bridge which is formed between conjugating cells of spirogyra, etc., and by which impregnation is effected.
Connate, united congenitally.
Connective, the end of the filament, between the anther sacs.
Connivent, converging.
Conopodium, a conical receptacle (used of flowers).
Contorted aestivation, used when the corolla appears spirally twisted, the petals being so arranged that one margin is external to a neighboring petal while the other is internal to the petal on the other side.
Contractile cells, in the anther, form a layer in its wall; their membranes are peculiarly thickened, and by their hygroscopic contractions the anther opens.
Convolute, applied to a leaf which is rolled up longitudinally in the bud.
Coppice, to cut back so as to produce shoots from old stumps.
Coralloid, resembling coral.
cordate, heart-shaped, as a leaf.
Corolla, the inner series of the floral envelope; the petals as a unit.
Corm, a bulb-like fleshy stem or base of a stem; a "solid bulb", as in crocus, colchicum, etc.
Corona, in narcissus, etc., a series of ligular structures on petals, which may be either free or united together. It gives the appearance of an additional floral whorl.
Corpuscle, a little mass of protoplasm which though imbedded in the general protoplasm of the cell is nevertheless an independent body, e.g. chlorophyll-corpuscle.
Corpusculum (of asclepiad pollinium), the little body connecting the pollen-masses and by means of which they become attached to insects.
Cortex, the portion of a stem or root external to the vascular tissues.
Corymbus, or Corymb, a flat-topped infloresence belonging to the centripetal or indefinite series.
Cosmic dust, the minutely divided inorganic particles suspended in the higher strata of the atmosphere; not necessarily of extra-terrestrial orgin.
Cosmopolitan plants are such as range almost over the entire globe; in contrast to plants that flourish only in a certain locality (endemic plants).
Costate, ribbed.
Cotyledon, the seed leaf or primary leaf or leaves in the embryo, containing stored food for the developing seedling and often resembling true leaves and manufacturing food.
Craspedromous, used of the lateral veins of a leaf which run undivided from midrib to margin.
Crateriform, goblet-or cup-shaped.
Crenate, said of a toothed leaf-margin, the teeth being rounded; scalloped.
Cross-fertilization, the fertilization of an egg-cell by a male cell borne on another individual; fertilization of the ovules of one flowers by the pollen from another individual. Occasionally used in error in the text for cross-pollination (which see). Many authors use the term as synonymous with cross-pollination, but the practice is not good.

GLOSSARY

Cross-pollination, the deposition on a stigma of pollen which has been brought from another flower. Cross-pollination, though probably leading to cross-fertilization, is not synonymous with this term.

Cruciferous, "cross-bearing", having cross-shaped flowers: used of the characteristically flowered family cruciferae.

Crustaceous, hard and brittle in texture; crustlike.

Cryptogamia, includes all plants exclusive of flowering plants: opposed to phanerogamia. An old term, persisting from times when the reproductive processes of these plants were less well-known than today.

Crystalloid, a crystal-like mass of proteid; a common form under which proteids are stored.

Cucullate, hooded, or resembling a hood.

Culmus, or Culm, the jointed and usually hollow stem of grasses and similar plants.

Cuneate, wedge-shaped.

Cupule, the bract-like cup which incloses the nut or nuts in many amentiferae; it is the husk of the hazel-nut, the cup of the acorn, the prickly envelope of the Spanish chestnut, etc.

Cusp, a sharp stiff point.

Cuspidate, sharp-pointed; ending in a cusp.

Cut, a term applied to the lobing of leaf-blades; incised; cleft.

Cuticle, a continuous film on the surface of a plant, formed of the cutinized outer surfaces of the epidermal cells.

Cutting, a severed vegetative or asexual part of a plant used in propagation.

Cutting test, cutting or otherwise opening seed for the purpose of determining their soundness or viability.

Cyma, or Cyme, a definite or centrifugal inflorescence: the laterals grow more strongly than the primary axis and overtop it.

Cyma composita, or compound cyme; a definite or centrifugal inflorescence, in which the ultimate parts (cymes) are also arrange in a cymose manner.

Cymose, arranged in cymes; cyme-like.

Cystolith, a concretion of carbonate of lime, generally deposited on a little tongue or peg of cellulose projecting into the cells of certain plants.

Cytoplasm, the protoplasmic body of a cell as opposed to the nucleus.

Daughter-cells, cells which arise by the division of any cell.

Deciduous, failing at maturity, not persistent.

Decompound, more than once-divided.

Decumbent, stems or branches in an inclined position, But the end ascending.

Decurrent, used of leaf-blades which have their bases extending downward along the stem.

Decussate, applied to leaves which are arranged in pairs alternately crossing each other at regular angles.

Definitive nucleus, the nucleus which is formed in the embryo-sac by the fusion of two, one from each end; the endosperm originates from it after fertilization has taken place.

Deflexed, turned abruptly downward.

Dehiscent, splitting open to discharge the contents, as a capsule or anther.

Dendritic, tree-like; repeatedly branched.

Denizen, an inhabitant, a plant belonging to a certain district. Strictly a plant resembling a native, but suspected of having been originally introduced.

Dentate, of leaf margins; toothed—the teeth pointing outwards, not forwards or backwards.

Dermatogen, the embryonic cellular layer at the apex of a stem or root from which the epidermis is developed.

Desmid, one of the conjugatae.

Dewing, to remove wings from seed.

Dextrorse, used of twining plants which turn from west through south to east, etc.

Diadelphous, stamens united into two sets.

Diadromous, having a fan-like arrangement of leaf-veins, as in gingko.

Dialypetalae, plants with petals separate from one another.

Diandria, the second class of Linnean system; includes all genera with perfect flowers having two stamens.

Diastase, a solid, white, soluble substance found in oats, potatoes, etc., after germination.

Diastole, used of the rhythmic expansion of a contractile cell or vacuole.

Diatom, a single organism inclosed in a bivalved siliceous test or frustule.

Diatomin, the brown pigment of diatoms.

Dichogamy, the maturing of pollen and stigma in a hermaphrodite flower at different times, to prevent self-fertilization.

Dicotyledon, plant with two seed-leaves or cotyledons.

Dictyodromous, or reticulate venation, are terms applied to lateral veins of leaves which break up into a network before reaching the margin.

Didynamia, the 14th class of the Linnean system, which includes flowers with four stamens, two long and two short.

Didynamous, applied to flowers having four stamens, one pair longer than the other.

Diffuse, loosely spreading.

Digitate, diverging, like the fingers spread.

Dimorphous, of two forms.

Dioecious, having staminate (male) and pistillate (female) flowers borne on different plants (as in Acer, Fraxinus, and Ilex).

Diosmosis, the transfusion of a fluid through imperceptible openings in a membrane.

Diptero-cecidia, gall-structures, due to dipterous insects.

Discoid, resembling a disc.

Discomycete, any fungus belonging to the group discomycetes, i.e. an asomycete in which the fruiting body is disc-shaped.

Discomycetous, pertaining to the group of fungi discomycetes.

Discopodium, a disc-shaped floral receptacle.

Disintegration, a resolution of a tissue into its constituent cells, or of any body into its constituents.

Disk, an enlargement or prolongation of the receptacle of a flower around the base of the pistil; the head of tubular flowers in compositae.

Displacement, in whorls, applied to the shifting of places of insertion of members, so that successive whorls are placed immediately above one another.

Dissected, divided into many segments or lobes.

Dissepiment, a partion-wall of an ovary or fruit.

Distichous, arranged in two rows.

Distinct, separate from each other; evident.

Divaricate, diverging at a wide angle.

Divergence, applied to the angle between the insertions of successive leaves on a stem.

Divided, used of leaf-blades to express the fact that they are deeply lobed.

Dormancy, continued suspension of growth or development in the presence of external conditions favorable for germination.

Dormancy, double, a combination of seed coat and internal dormancy.

Dormancy, embryo, dormancy due to internal conditions of the embryo.

Dormancy, internal, dormancy due to internal conditions of the stored food or embryo.

Dormancy, seed coat, dormancy due to a seed coat impermeable to water or oxygen.

Dormancy, secondary, suspension of growth or development after original dormancy has been broken.

Dormant eyes or buds, or reserve-buds, are buds which arise in the leaf-axils in the usual way, but which do not forthwith expand into shoots; they remain—often many years—until stimulated into activity by some special event.

Drupe, a stone fruit, or fleshy nonsplitting fruit with a bony inner layer (endocarp) and usually one-seeded (as in chionanthus, cornus, and prunus).

Duct, a continuous tube, arising either by the running together of cells (fusion), or by the separation of cells, when it is lacunar in nature; a canal formed by a row of cells having lost their partitions.

Dwarf-male, of oedogonium; the little few-celled plant arising from an androspore which gives rise to the spermatozoids. It is formed adjacent to the oogonium.

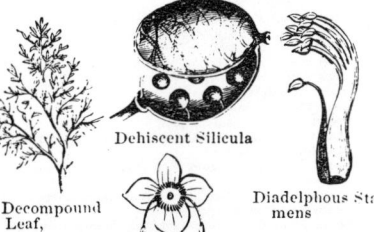

Decompound Leaf. Dehiscent Silicula. Diadelphous Stamens. Diandrous Flower.

Ecotype, a race or subspecies naturally selected on the basis of a certain local habitat and climate.

Ectoplasm, the pellicle-like outmost layer of protoplasm in a cell. It is clear and hyaline, and less fluid than the endoplasm.

Egg-cell, or Ovum; the female generative cell.

Ellipsoid, a solid body elliptic in the longitudinal section.

Embryo, the rudimentary plant within the seed; sometimes called germ.

Embryo-cell, the cell borne at the distal end of the suspensor, which gives rise to the embryo, or to the greater part of it.

Embryo, immature, an embryo which is not fully developed or capable of germination when the seed is harvested.

Embryo-sac, the large cell in the nucellus of an ovule, in which the eg-cell, and ultimately the embryo, arises.

Endemic, restricted to a given region or locality.

Endocarp, the inner layer of the pericarp (for example, the bony part or stone of the fruit in prunus).

Endophytic, living within the tissues of another plant, though not necessarily parasitic upon them.

Endoplasm, the soft, inner granular protoplasm of a cell.

Endosmosis, the transmission of fluids through porous membranes from the exterior to the interior.

Endosperm, the nutritive tissue of seeds, in which the embryo is imbedded.

Endospores, asexual reproductive cells produced inside the original cells in bacteria.

Endothecium, in flowering plants, the layers of the wall of the anther internal to the exothecium.

Ennobling, the art of transferring a branch or bud of one plant to another, and causing them to unite.

Entire, untoothed: applied to the leaf-margin, petals, etc.

Entomophilous plants, such as have flowers pollinated by insect agency.

Enzyme, any of the unorganized ferments which exist in seeds, as diastase, pepsin, etc.

Ephemeral, applied to flowers which endure only for a few hours or for a day; opening but once.

Epicotyl, the portion of a plant above the cotyledons; restricted to embryos and seedlings.

Epidermis, that layer of cells which forms the enveloping mantle of multicellular plant-bodies. It may be replaced in perennial plants by cork.

Epigeal, growing above the ground.

Epiphragm, of mosses, the membrane remaining after the fall of the operculum, stretched across the mouth of the capsule in polytrichaceae.

Epiphyllous, applied to structures growing on leaves.

Epiphytes, plants growing attached to other plants (or animals), but not parasitically.

Epiphytic, growing on other plants, but not parasitic.

Erythrophyll, a red sap-pigment frequent in foliage leaves, especially in autumn.

Ethereal oils, oils of wide occurrence in plants, and of various chemical composition; to the presence of these ethereal or volatile oils are due most of the odors of plants.

Evanescent, early disappearing.

Evergreen, bearing green leaves throughout the year.

Evolute, turned back.

Exalbuminous, applied to seeds which are destitute of endosperm or perisperm, the food-material being stored in the embryo itself.

Excoriation, of glandular hairs; applied to the act of throwing off the cuticle as a blister.

Excurrent, with a tip projecting beyond the main part of the organ.

Exfoliate, to come away in scales or flakes, as the bark of a tree.

Exocarp, the outer layer of the pericarp.

Exogamy, the tendency often exhibited by closely related gametes to prevent pairing.

Exogenous, forming new tissue outside the older.

Exosmosis, the passage from within outwards of fluids through a membrane.

Exothecium, the outmost layer or epidermis of an anther.

Exserted, prolonged past surrounding organs.

Estipulate, without stipules: often used (though erroneously) in cases where the stipules are early deciduous.

Extine, the outer coat or membrane-layer of a pollen grain. It is, however, internal to the perine.

Extraction factor, the weight of cleaned seed per given weight of fresh fruits, usually expressed in percent.

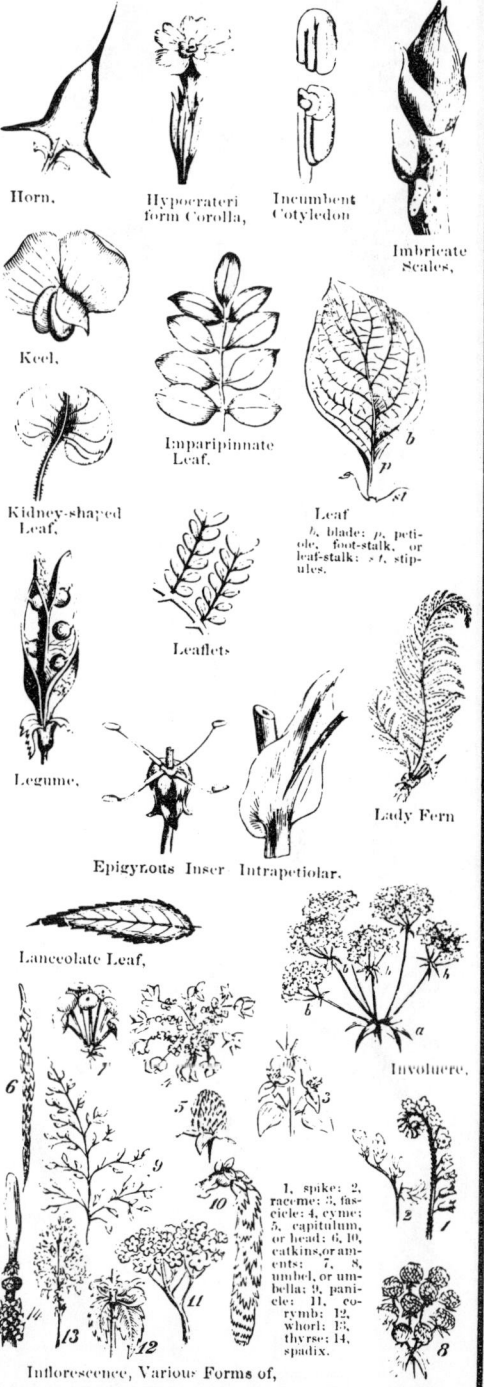

Horn. Hypocrateriform Corolla. Incumbent Cotyledon. Imbricate Scales. Keel. Imparipinnate Leaf. Leaf: b, blade; p, petiole, foot-stalk, or leaf-stalk; st, stipules. Leaflets. Legume. Lady Fern. Epigynous Inser. Intrapetiolar. Lanceolate Leaf. Involucre. 1, spike; 2, raceme; 3, fascicle; 4, cyme; 5, capitulum, or head; 6, 10, catkins, or aments; 7, 8, umbel, or umbella; 9, panicle; 11, corymb; 12, whorl; 13, thyrse; 14, spadix. Inflorescence, Various Forms of.

GLOSSARY

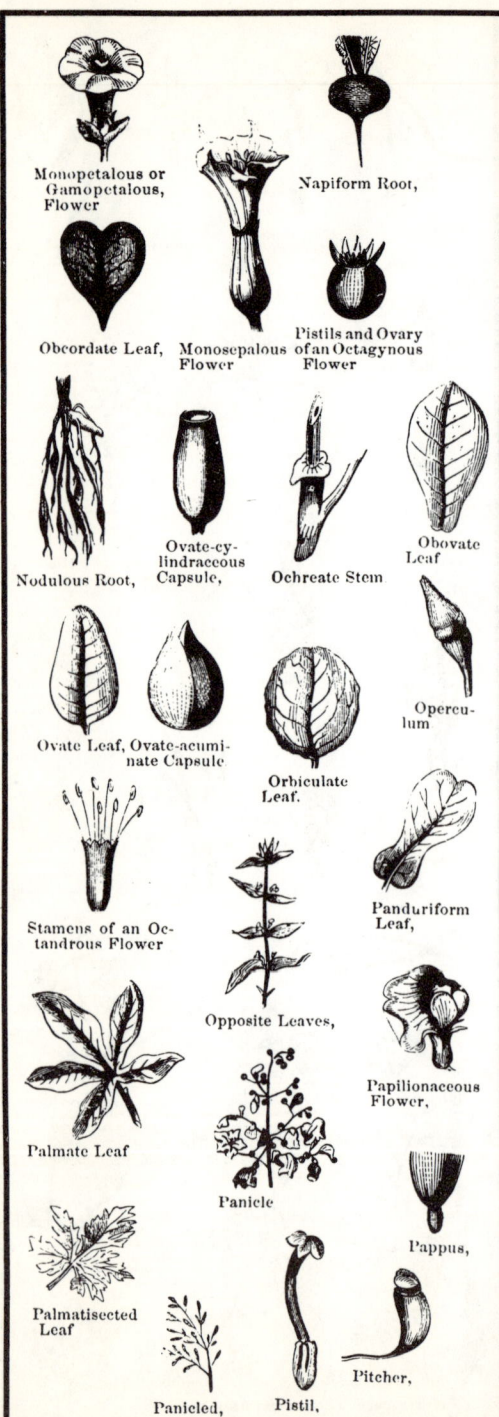

Monopetalous or Gamopetalous, Flower
Napiform Root,
Obcordate Leaf, Monosepalous Flower
Pistils and Ovary of an Octagynous Flower
Nodulous Root,
Ovate-cylindraceous Capsule,
Ochreate Stem
Obovate Leaf
Ovate Leaf, Ovate-acuminate Capsule
Operculum
Orbiculate Leaf.
Stamens of an Octandrous Flower
Panduriform Leaf,
Opposite Leaves,
Papilionaceous Flower,
Palmate Leaf
Panicle
Pappus,
Palmatisected Leaf
Paniceled,
Pistil,
Pitcher,

Extravasation, an escape from the proper vessels into surrounding tissues: used of fluids.
Extrorse, applied to such anthers as open towards the outer whorls of a flower, i.e. away from the gynoeceum.
Eye, of potato, etc, an undeveloped bud.
Eye-spot, in motile gametes and spermatozoids, a little red pigment-body contained usually in the anterior extremity, and supposed to be sensitive to light.

Fairy-ring, a phenomenon observed in meadows, and due to the growth of certain fungi.
Fasciation, used of monstrous expansions of stems, which resemble several stems fused together in one plane.
Fascicula, or fascicle, a dense cluster of flowers, leaves, roots, etc.
Fastigiate, stems or branches which are nearly erect and close together.
Ferment, a substance produced by the protoplasm, which induces chemical change or fermentation in some substance without itself etering into or being affected by the process.
Fertilization, the union of a sperm or male nucleus and an egg or female nucleus within an ovule, resulting in the development of an embryo plant within a seed.
Fertilizing-tube, in peronospora, the tubular outgrowth of the antheridium which penetrates the oogonial wall and by which the male substance passes to the egg-cell.
Fibre, any delicate filament; also, a thick-walled tapering cell.
Fibrillose, with fibres or fibre-like organs.
Fibrils of nucleus; the segments into which the nuclear reticulum breaks up at division; they are also termed chromosomes.
Fibrous layer, of anther: the specially thickened portion of the wall which brings about dehiscence.
Filament, the stalk of an anther.
Filiform, slender, thread-like.
Fimbriate, fringed by fine subdivision of the margin; having fine, hair-like marginal processes.
Fistular, hollow, reed-like.
Flagellum, the whip-like process or filament of protoplasm which serves as an organ of motility; also a shoot sent out from the bottom of a stem, as in the strawberry; a runner.
Floccose, composed of or bearing soft hairs or wool.
Flora, the aggregate of the plant-population of any district; also, the term given to a systematic description of the same.
Floral, belonging to the flower.
Floret, a small flower in a cluster or in a compact inflorescence, as in the compositae.
Flower, in phanerogams the growth which comprises the reproductive organs and their envelopes; a shoot modified for the production of spores (pollen-grains and embryo-sacs).
Flower, bisexual, a flower having both stamens (male element) and pistil (female element).
Flower, imperfect, a flower having either stamens or pistil but not both.
Flower, perfect, a flower having both stamens and pistil; bisexual.
Flower, pistillate, a flower having a pistil but no stamens; a female flower.
Flower, staminate, a flower having stamens but no pistil; male or pollen-producing flower.
Flower, unisexual, a flower having organs of a single sex, either stamens or pistil.
Flowering glume, the outer of the two chaffy scales inclosing the several flowers of a grass; it is frequently awned.
Folium fulcrans, the subtending leaf of a flower; a bract.
Follicle, a dry-one-celled fruit which opens along one side only (as in magnolia).
Foot, the sucker by means of which a young fern-plant is temporarily attached to the prothallium.
Foreign, applied to pollen from another flower.
Form, a subdivision of a botanical species or variety distinguished by some minor character and designated by the abbreviation "f." (forma) preceding the Latin name. See variety.
Free, separate from other organs; not adnate.
Frond, the leaves of ferns.
Frugivora, animals which live upon fruits.

Fruit, the ripened ovary of a flower, containing the seeds and composed of the usually thickened ovary wall (pericarp) and any other closely associated parts.
Fungus, a cellular cryptogam, distinguished for its want of chorophyll; it is either saprophytic or parasitic.
Funicle, the stalk of an ovule or seed; a funiculus.
Furfuraceous, scurfy; covered with bran-like scales.
Fusiform, spindle-shaped: applied to roots, etc. which taper both ways from the middle, as the radish.

Galeate, helmet-shaped; having a galea or helmet.
Gall, a vegetable excrescence produced by the deposit of the egg of an insect in the bark or leaves of a plant; a hypertrophied growth due to some irritating cause.
Gametangia, cells from which gametes are developed.
Gamete, a sexual cell.
Gametophyte, the sexual generation of plants.
Gamopetalous, with petals more or less united.
Geitonogamy, crossing between separate flowers growing on the same plant.
Gemma, a small undeveloped shoot; a shoot-bud.
Gemmation, the act or process of budding.
Generative cell, in pollen-grains, that cell which ultimately fertilizes the egg-cell.
Genetic spiral, the spiral line passing through the point of insertion of equivalent lateral members (leaves) on an axis in order of age from older to younger.
Genus, an assemblage of species; its name, together with that of the species, gives the name to the plant.
Geotropism, applied to the power or tendency of some plants to grow towards the earth.
Germen, the ovary.
Germination, the development of the seedling from the seed; sprouting.
Germination, epigeous, the usual type of germination in which the cotyledons are brought above the ground.
Germination, hypogeous, a type of germination in which the cotyledons remain below ground (as in juglans, quercus, and torreya).
Germination percent, the percent of a given number of seeds germinating (usually in a given time).
Germination, potential, number of seeds germinating plus the number of sound seeds ungerminated at the close of the test, expressed in percent of the total number of seeds tested.
Germination, real, percent of sound seed germinating.
Germination tests, indirect, tests to determine the viability of seeds without actually germinating them, such as cutting tests, staining with chemicals, excising embryos, and the like.
Germinative capacity, the percent of seed actually germinating, regardless of time.
Germinative energy, the percent of seed germinated at the time the trend of germination reaches its peak.
Germinator, an apparatus in which seeds are tested for germination.
Germinator, standard, an apparatus for testing seed germination. It consists of a porous surface (porous clay, plaster of paris, blotting paper, or cloth) on an impervious surface, is kept continuously moist by wicks or similar arrangements, and usually is covered by an impervious transparent object such as a bell jar or inverted glass funnel. (In general use are Geneva, Jacobsen, Rodewald, Stainer, Toumey, and some other germinators.)
Gills, the radiating plates on which the basidiospores of agarics are produced.
Glabrous, smooth; not hairy.
Glaucous, covered with a bluish or whitish bloom.
Gleba, the chambered, sporogenous layer of a gasteromycetous fungus.
Globoid, the tiny mass of magnesium and calcium phosphate which is often present in aleurone grains (which see).
Globose, spherical in form; globe-shaped.
Glomerate, in a compact cluster.
Glomerule, a cymose inflorescence formed into a head, as in the globe-thistle.
Glucoside, a compound consisting of glucose and an aromatic body.
Glumes, the chaffy, bract-like scales on the inflorescences of grasses and sedges.
Goneoclinic, used of hybrids which approximate to one or other parent-form rather than standing midway between them.

Graft-hybrid, a hybrid supposed to have arisen by budding or grafting.
Gymnosperm, a member of the gymnosperms, one of the two groups of seed plants and characterized by naked seeds borne usually on scales of a cone, or sometimes singly. See angiosperm.
Gynandria, the 20th class of the Linnean system.
Gynobase, a prolongation or enlargement of the receptacle, supporting the ovary.
Gynoeceum, the carpel, or aggregate of carpels, in a flower; the female portion of a flower as a whole.

Habitat, the natural abode of a plant.
Haematochrome, the red pigment found in the eyespots of chlamydomonadeae and zoospores.
Halophytes, plants which flourish on soils rich in salt; saltworts.
Haulm, the stalk of a grass of any kind.
Haustoria, the specialized roots of parasites.
Haustorium, the sucker of a parasitic plant.
Head, flower, a dense, rounded flower cluster with the individual flowers stalkless or nearly so (as in baccharis and cephalanthus).
Herbaceous, of the color, texture, etc., of a herb.
Herbal, a book of descriptions of plants with especial reference to their medicinal properties; herbals were usually copiously illustrated.
Herbarium, a collection of dried plants systematically arranged. (Formerly it signified an illustrated herbal.)
Hermaphrodite, applied to a flower which has both stamens and carpels.
Heteroecism, the act of passing through different stages of development on different hosts; as in fungi.
Heterogamous, applied to plants that bear two kinds of flowers which differ sexually.
Heterogamy, the state or quality of being heterogamous (which see); cross-pollination.
Heteromorphism, here used to designate the various modifications of equivalent members in connection with different functions, analogous to that existing among the polyps of a coral.
Heterophyllous, bearing leaves of more than one form on the same stem; applied especially in respect of foliage-leaves.
Heterosporous, having spores of different kinds, especially macrospores and microspores.
Heterostyled, when the flowers of a plant differ in the relative length of their styles: opposed to homostyled.
Hilum, the scar or point of attachment of the seed.
Hirsute, bearing rather stiff hairs.
Holosericeus, covered with fine silky hairs.
Homosporous, having spores all of a kind.
Hortus vivus, an old term for a dried collection of plants, now called a herbarium (hortus siccus is also used in the same sense).
Hull, the outer covering of a fruit, especially if smooth or relatively so, as distinguished from the shell (as in carya or juglans).
Humus, vegetable mould; a soil largely composed of decaying vegetable matter.
Husk, an outside envelope of a fruit, especially if coarse, harsh, or rough (as in corylus).
Hybrid, a plant resulting from a cross between two or more parents that are more or less unlike.
Hybridization, the act of crossing different species and so producing hybrids.
Hydrophytes, plants which live in water.
Hydrotropism, the particular irritability of plant members (especially roots) whereby they respond by curvatures to moisture in the environment, turning towards or away from it.
Hymenium, hymenial layer; the spore-bearing surface of a fungal receptacle.
Hypanthium, a term given to any special enlargement of the receptacle, as in the rose.
Hypha, the filamentous element of the thallus of a fungus.
Hyphodromous, used when the veins of a leaf run so that they are not visible on the surface.
Hypocotyl, the stalk or stem of the embryo or seeding between the attachment of the cotyledons and the radicle or root.
Hypocrateriform, salver-shaped: used of corollas, etc., which are tubular below and suddenly expand into a flat limb.
Hypogeal, underground; growing beneath the surface of the earth.

188

GLOSSARY

Imbricating, overlapping like the tiles of a roof.
Imperfect, flowers with either stamens or pistils, not with both.
Incised, of leaves, cut irregularly and sharply.
Included, not projecting beyond surrounding parts.
Incumbent, with the back against the hypocotyl.
Indumentum, a hairy covering or coating.
Indusium, the scale-like outgrowth of a fern leaf enveloping the sorus.
Inferior, (1) of the ovary; adherent to the calyx; (2) of the calyx, free from the ovary; (3) in regard to the relation of parts of flower to the axis; farthest from the axis.
Inflorescence, the mode of branching of the flower-bearing part of a plant; or, the actual cluster of flowers (the common use of the term).
Infundibuliform, infundibular, funnel-shaped.
Innovatio, a new-formed shoot.
Insectivorous plants, plants which catch insects and absorb their juices.
Integument, the envelope—single or double—of an ovule.
Internode, the portion of a stem between the points of insertion of leaves.
Intine, the internal layer of the wall of a pollen-grain.
Introrse, of the anther; dehiscing towards the center of the flower.
Intussusception, the taking up by a living organism of new particles between those already in existence.
Invertin, a ferment which converts cane-sugar into glucose.Involucral, appertaining to the involucre.
Involucre, a whorl of bracts surrounding a flower cluster or a single flower or the fruits developed there-from.Involute, rolled inward.
Irregular, a flower in which one or more of the organs of the same series are unlike.
Isogametes, equivalent gametes or sexual cells.
Isoplanogametes, in algae; motile sexual cells which are equal in size.

Kernel, the inner, often edible portion of a nut. Also, the portion of the seed which contains the embryo.

Labellum, the median member of the inner perianth-whorl in orchids.
Laciniated, cut into narrow lobes.
Lacuna, a space, especially an intercellular space, originating by the separation or breaking down of cells.
Lamella, a thin plate as in the gills of agarics. See gills.
Lanceolate, shaped like a lance-head; narrower than oblong, and tapering towards the apex.
Latex, plant juice, often a milky juice.
Laticiferous, containing latex.
Layer, a stem or branch which takes root while still attached to the parent plant and tends eventually to become a separate individual plant.
Leaf-axil, the angle formed by a leaf and the portion of stem immediately above its point of insertion.
Leaves, laterally developed members of limited growth, which spring in geometrical succession from the outer layers of tissue below the growing point of the stem.
Legitimate union, in heterostyled flowers.
Legume, or Pod; a monocarpellary fruit dehiscing down both sutures.
Lepidote, -us, beset with scrufy scales.
Liane, liana, a climbing plant with a woody, perennial stem.
Libriform cells, strong, spindle-shaped cells with inconspicuous pittings, thick walls, and usually destitute of protoplasmic contents. They occur in wood.
Lichen, an organism compounded of a fungus and an alga living together symbiotically.
Lignin, an aromatic substance (or number of substances) present in the membrane of woody tissue. To it are due the characteristic properties of wood.
Ligulate, provided with a ligule.
Ligule, Ligula, (1) the thin scarious projection from the summit of the leaf-sheath in grasses; (2) the corolla of a ray-floret in the compositae; (3) a tongue-like outgrowth on the leaf met with in selaginella and isoetes just above the insertion of the sporangium.

Limb, the expanded part of a petal, sepal, or gamopetalous corolla.
Linear, several times narrower than long, with the margins parallel.
Linear-lanceolate, intermediate in form between linear and lanceolate.
Lines of vegetation, for any species, are the lines obtained by joining all the places in a given directic at which that species is checked in its distributic by climatic or other conditions; the resultant fi obtained by joining all the lines of vegetation c the distribution of the species in question, and be termed the line of distribution.
Lithophytes, plants which grow on stones, and deri their nutriment in the main from the atmosphere.
Liverwort, a term applied to any member of the hepaticae.
Lobe, any division of an organ; a rounded projection or division.
Lodicules, tiny scales, usually two in number, which occur in the flowers of grasses, and are supposed to represent the perianth.
Lomentum, a legume which separates into 1-seeded articulations or joints.
Locuticidal, applied to capsules which split longitudinally.
Lodicules, minute hyaline scales subtending the flower in grasses.
Lunate, crescent-shaped.
Lyrate, pinnatifid, with the terminal lobe or segment considerably larger than the others.

Macerator, a special apparatus for separating from fruits, either fleshy or dry.
Macropodous, applied to embryos in which the hypocotyl is enormously enlarged, constituting the greater part of the embryo.
Macrospores, used of the larger (so-called female) spores of heterosporous plants: opposed to microspores.
Manubrium, the cell in the antheridium of Characeae which projects inwards from the shield, and ultimately bears the antheridial filaments.
Medulla, pith.
Megagametes, used of the larger, and presumably female, motile sexual cells of certain algae.
Melliferous, honey-bearing.
Membranous, thin, pliable, and rather soft; of the texture of a membrane. Syn. membranaceous.
Mericarp, one of the achene-like fragments into which a syncarpous, polycarpellary fruit (schizocarp) breaks up. Used especially of umbelliferae.
Meristem, embryonic tissue: growing cell tissue at the ends of young stems, roots, etc.
Mesophyll; the whole of the internal ground-tissue of a leaf-blade.
Metabolism, the chemical changes which take place in the protoplasm and which it causes in other substances; the phenomena resulting from chemical changes in the protoplasm.
Micellae, name given to molecular aggregates, just as molecule is the name given to atomic aggregates.
Microgametes, used of the smaller, presumably male, motile sexual cells of certain algae.
Micro-millimeter, the one-thousandth part of a millimeter.
Micropyle, the pore of the ovule through which the pollen tube enters, and the corresponding scar in the seed.
Microspores, the smaller or so-called male spores of heterosporous plants: opposed to macrospores.
Midrib, the central or main vascular bundle of a leaf.
Monadelphous, when the stamens are all united together by their filaments into a tube or column.
Monandria, the 1st class of the Linnean system.
Moniliform, like a necklace or string of beads.
Monocarpellary, consisting of one carpel.
Monocotyledonous, having only a single cotyledon or seed-leaf.
Monoecious, having staminate (male) and pistillate (female) flowers borne separately on the same plant (as in betula, pinus, and quercus).
Morphology, that department of botanical study which deals with the form of the plant body, including its development, the growth of its distinct members, etc.
Mother-plant, that parent of a hybrid upon which the seed is matured.
Mucronate, with a short sharp abrupt tip.

Mucronulate, diminutive of mucronate.
Muricate, roughened with short hard processes.
Mulicous, pointless, or blunt.
Mycelium, the filamentous vegetative body of a fungus.
Myco-cecidium, a gall which owes its origin to the attacks of fungi.
Mycorhiza, a root invested by a fungal mantle; supposed to be a case of symbiosis.
Mycosis, a diseased condition of animal tissues alleged to be due to the presence of a mould-fungus.
Myrmecophilous, used of plants which attract ants, the latter often living altogether upon the plant and affording it protection against certain enemies.

Naturalized, plants not indigenous to the region, but so firmly established as to have become part of the flora.
Nectary, a honey-secreting gland or part of a flower.
Neroli, oil of, the ethereal oil yielded by the flowers of the orange tree.
Neuter flowers, flowers destitute of functional stamens or carpels.
Node, the part of a stem at which a leaf or whorl of leaves is inserted.
Nodose, or Nodosus, knotty; having well-marked nodes or knots.
Nodulated, having small knots: diminutive of nodose.
Nucellus, a tissue composing the central part of the young ovule.
Nuclear plate, the assemblage of nuclear fibrils in the equator of a nucleus during the division of the latter.
Nucleus, the central denser structure of a cell. Also the kernel of a seed.
Nut, a nonsplitting one-seeded fruit, with hard woody shell (as in carya and corylus). In common usage, a hard-shelled fruit or seed containing an edible kernel.
Nutlet, a small nut or nutlike fruit or seed (as in alnus, carpinus, and ostraya).

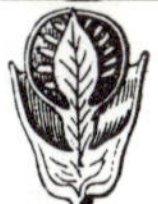
Ovary

Obcordate, inversely heart-shaped.
Oblanceolate, inverse of lanceolate.
Oblong, longer than broad with the sides nearly parallel, or somewhat curving.
Obovate, ovate with the broader end at the apex.
Obovoid, inversely egg-shaped, with the border end uppermost.
Omphalodium, the scar at the hilum of a seed.
Ontogeny, the history of the individual development of an organized being.
Oogonium, the cell in which the female sexual cell or cells are produced; especially amongst thallophytes.
Oophyte, that stage in the life-cycle of a plant which bears the sexual organs.
Oospore, a fertilized egg-cell.
Operculate, with an operculum.
Operculum, the lid of a moss capsule.
Orbicular, approximately circular in outline.
Order, a division of plants intermediate between class and genus, consisting usually of a group of genera related to one another by structural characters common to all.
Orthostichies, vertical ranks of leaves.
Orthotropous, applied to an ovule with straight nucellus wherein the micropyle is at a point far removed from the funicle.
Osmosis, the tendency of fluids to pass through porous membranes; the phenomena attending the passage of fluids through porous membranes.
Ostiole, the aperture of the conceptacle in the fucaceae.
Ovary, the part of the pistil that contains the ovules or immature seeds; the closed chamber-like portion of a single free carpel, or the many chambers of several united carpels in which the ovules are produced.
Ovoid, oval or egg-shaped.
Ovule, the body which after fertilization becomes the seed.
Ovuliferous scale, the ovule-bearing scale of conifers.

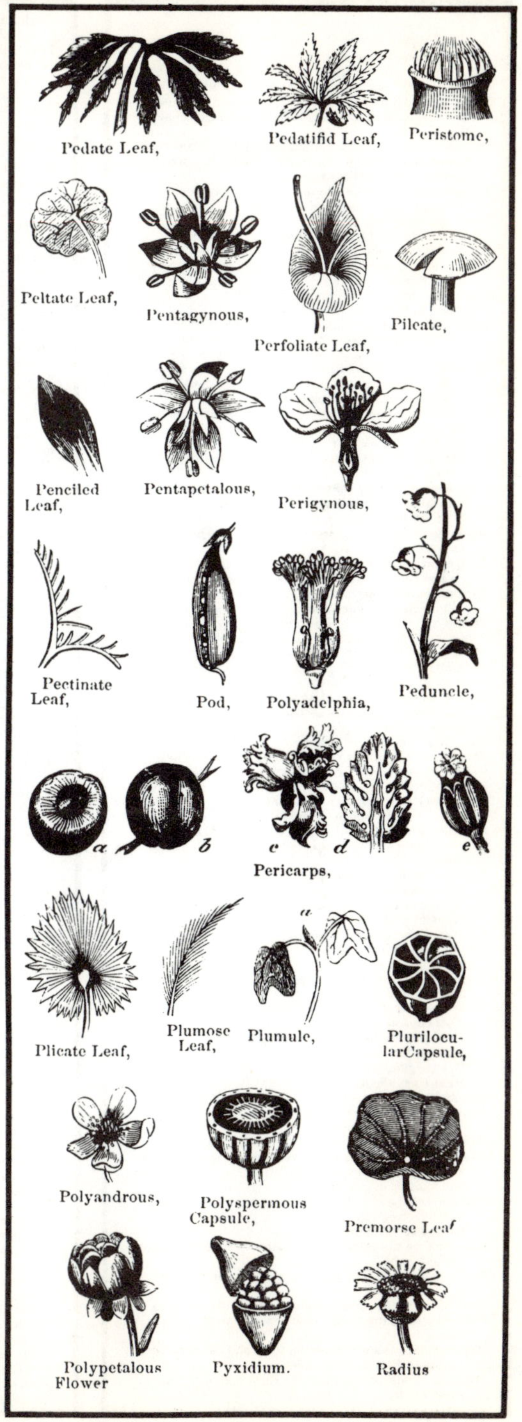
Pedate Leaf,
Pedatifid Leaf,
Peristome,
Peltate Leaf,
Pentagynous,
Perfoliate Leaf,
Pileate,
Penciled Leaf,
Pentapetalous,
Perigynous,
Pectinate Leaf,
Pod,
Polyadelphia,
Peduncle,
a
b
c
d
e
Pericarps,
Plicate Leaf,
Plumose Leaf,
a
Plumule,
Plurilocu-larCapsule,
Polyandrous,
Polyspermous Capsule,
Premorse Leaf
Polypetalous Flower
Pyxidium.
Radius

GLOSSARY

Palaeo-botany, fossil botany.
Palate, a projection in the throat of a personate corolla (or corolla such as that of the snapdragon).
Palea, the inmost of the glumes which inclose the individual flowers of grasses; a chaffy scale or chaff-like bract.
Palisade-cells, the green assimilating tissue, consisting of cylindrical cells, usually found towards the upper surface of the leaf-blade.
Palmate (of leaf-blades), lobed so that the projections radiate from the point of insertion.
Panicle, a compound raceme, or an open and branched flower cluster (as in aesculus).
Papilionaceous, like a butterfly: a term applied to the corolla of a section of Leguminosae, including the pea and bean, etc.
Papilla, a minute nipple-shaped projection.
Papillose, bearing papillae.
Pappus, the modified calyx-limb in the composite family (compositae), forming a crown of bristles, awns, or scales (as in baccharis and chrysothamnus).
Paraphyses, sterile filaments accompanying the sexual organs in mosses, the asci and basidia of basidiomycetes, and in other cases.
Parasite, a plant which lives upon and obtains organic nutriment from the tissues of a living plant (or animal).
Parastichies, secondary spirals in the arrangement of leaves.
Parenchyma, usually thin-walled tissue consisting of cubical or polygonal cells, and forming the pulp of leaves, fruits, etc.
Parthenocarpy, the development of fruit without fertilization.
Parthenogenesis, the development of an egg-cell into an embryo without fertilization taking place.
Parthenogonidia, certain reproductive cells in a volvox-colony which propagate the plant asexually.
Partite, cleft, but not quite to the base.
Patelliform, disc-shaped; circular with a rim.
Pedicel, the stalk of each individual flower in a cluster.
Peduncle, the stalk of a flower cluster; also used for the stalk of a solitary flower.
Peltate, shield-like: said of leaves when the petiole is attached to the under surface of the blade and not to the margin.
Pendulous, hanging, or drooping.
Penicillate, with a tuft of hairs or hair-like branches.
Perfect, flowers with both stamens and pistils.
Periblem, the embryonic tissue at a growing point from which the primary cortex arises.
Pericarp, the wall of the ripened ovary, or fruit.
Perichaetium, the sheathing structures in muscineae which envelop the clusters of archegonia and antheridia.
Peridium, the outer envelope investing the fructification in certain fungi.
Perine, the outmost layer of sculpturing on the membrane of pollen grains.
Perisperm, the tissue of the nucellus, in which, sometimes, food material is stored for the ultimate use of the embryo. It is external to the embryo-sac. In many old systematic books it is used for all food material of seeds which is external to the embryo.
Peristome, the ring of teeth around the mouth of a moss capsule.
Perithecium, the flask-shaped cavity in which asci are produced in certain fungi.
Petal, a corolla leaf.
Petaloid, -ine, like a petal.
Petiole, the stalk of a leaf.
Petit grain, name for the ethereal oil yielded by the leaves of the orange tree.
Phanerogamia, seed-bearing or flowering plants.
Phenology, the science of the relations between climate and periodic biological phenomena, as the flowering and fruiting of plants.
Phloem, soft bast; the soft outer portion of a vascular bundle, of which sieve-tubes are the most characteristic constituents.
Phrygana, an old term for a growth of stiff and prickly under shrubs.
Phycocyanin, the blue pigment of the cyanophyceae or lowest algae.
Phycophaein, the brown pigment of the brown algae.
Phycoerythrin, the purple coloring matter of red seaweeds.
Phylloclade, a branch assuming form and function of a foliage-leaf: same as cladode.
Phyllode, a petiole assuming the form and function of a leaf blade.
Phyllotaxis, leaf arrangement; the arrangement or order of distribution of leaves on the stem.
Phylogeny, or Phylogenesis, the history of the genealogical development of an organized being; the race history of an animal or plant, as distinguished from ontogeny, the history of individual development.
Phylum, a main division of the vegetable kingdom.
Pileus, the cap-shaped receptacle of a basidiomycetous fungus; the umbrella-like part of a mushroom.
Pili fasciculati, tufted hairs.
Pili stellati, stellate hairs.
Pilose, with long soft hairs.
Pinnate, when leaflets are arranged on either side of a common rachis or petiole.
Pinna, a primary division of a pinnately compound leaf.
Pinnatifid, pinnately cleft to the middle or beyond.
Pinnule, a division of a pinna.
Pistil, the female or seed-producing organ of a flower, composed of stigma, style, and ovary and derived from one or more modified floral leaves (carpels).
Pistillate, female; bearing pistils or seed-producing organs but no stamens.
Pitcher, a tubular or excavated leaf, usually containing a liquid; an ascidium.
Pith, the central cellular part of a stem or root.
Pits, thin places or depressions on cell-walls.
Placenta, the part of the carpel which bears the ovules; in vascular cryptogams, the portion of leaf surface bearing the sporangia.
Plaited, folded; folded into plaits lengthwise; plicate.
Plant formation, a term used to indicate the presence of two or more types of plant community intermingled together often in obvious strata.
Plasmodium, in the myxomycetes; a mass of naked multi-nucleate protoplasm exhibiting amoeboid movements.
Pleomorphism, the occurrence of more than one independent form in the life cycle of a species, especially in fungi and bacteria.
Pleurocarpous, used of mosses in which the archegonia are borne, not at the tips of the main but of secondary shoots.
Plumule, the bud or growing point of the embryo, which develops into the stem and leaves. Syn. Epicotyl.
Podium, a term for the torus or floral receptacle.
Polar nuclei, the two nuclei—one from each end of the embryo-sac of angiosperms—which approach one another and fuse to form the definitive nucleus of the embryo-sac.
Pollarding, the act of removing the crown of a tree so as to induce it to throw out branches around the place of amputation.
Pollen, the mass of fecundating cells or grains contained in the anther.
Pollen-grain, one of the fecundating cells of the pollen; the microspore in flowering plants.
Pollen-sac, the sporangium in which the microspores or pollen-grains of flowering plants are developed.
Pollen-tube, the tubular outgrowth of a pollen-grain by means of which fertilization is achieved.
Pollinia, masses of coherent pollen-grains.
Polycarpellary, having or consisting of a number of carpels.
Polychotomous, branching repeatedly into equivalent portions.
Polyembryony, the production of more than one embryo in a single seed.
Polygamous, having both perfect flowers and unisexual, or imperfect (i.e., staminate and pistillate) flowers.
Polygamo-dioecious, having both perfect and unisexual flowers on the same plant, the staminate flowers and pistillate flowers on different plants.
Polygamo-monoecious, having perfect flowers and the two kinds of unisexual flowers (staminate and pistillate) all on the same plant.
Polyhedra, angular bodies which arise from the zoospores into which the zygote of hydrodictyon breaks up. Ordinary hydrodictyon-nets arise inside them.
Pome, a fleshy fruit derived from several carpels and formed in part by the fleshy receptacle (as in the apple, pear, and some others of the rose family).
Porogamic, used of flowering plants in which the pollen-tube effects an entrance to the ovule by the micropyle.
Porous, used of dehiscence of anthers, etc., by means of holes.
Prechilling, a method of pretreatment to overcome dormancy in which seed are held at near-freezing temperatures and on a moist medium.
Pregermination, see stratification.
Prickle, a sharp-pointed process of the epidermis or cortex, but destitute of vascular tissue.
Primordial utricle, that portion of the cell-protoplasm which forms a bag in contact with the cell-wall. An old name which has persisted in the terminology.
Procumbent, lying along the ground.
Prolepsis, something of the nature of an anticipation.
Pro-mycelium, the limited tubular growth arising from the chlamydospores in hemibasidii and uredineae, from which conidia are abstricted.
Prophylla, bractlets.
Prostrate, lying flat on the ground.
Protandrous, proterandrous, used of flowers when the anthers dehisce before the stigmas are receptive.
Proteid, a nitrogenous substance of complex constitution, generally of a viscid nature and rarely crystallizable. The proteids include albumin, globulin, peptone, etc.
Prothallus, -ium, the structure produced by the germination of the spore of ferns, bearing sexual organs, and from which the young plant arises and derives nourishment for a time; also the homologue of this in flowering plants.
Protogynous, proterogynous, used of flowers in which the stigmas are receptive before the pollen of the same flower is discharged.
Protonema, the filamentous growth of a moss from which the leafy shoots arise by budding.
Protoplasm, the living and formative organic substance of plants and animals; living matter in its simplest form, serving as the basis of both animals and plants, and consisting of carbon, oxygen, hydrogen, and nitrogen, colorless, transparent or nearly so, and somewhat viscid in consistence.
Protoplast, the protoplasmic cell-body; a simple one-celled organism.
Provenience (or provenance), the geographical source or place of origin of a lot of seed.
Pseudo-hermaphrodite flowers are such as have been functionally unisexual by the suppressing of either stamens or carpels.
Pseudomorph, a term borrowed from mineralogy; an unusual or altered form.
Puberulent, with very short hairs.
Pubescent, with hairs.
Pulverulent, powdery.
Pulvinate, cushion-like.
Pulvinus, the enlargement of a petiole or leaf-stalk at its point of insertion on the stem, or of a secondary petiole at its point of insertion on the leaf-rachis.
Purity, percent by weight of clean, whole seed, true to species, in a sample of mixed impurities and seed.
Pycnidium, in fungi; a receptacle or cavity of varying form, in which conidia (pyco-conidia) are produced: especially in ascomycetes.
Pyrenoids, refractive bodies imbedded in the chlorophyll of many green algae.
Pyriform, pear-shaped.

Race, a group of plants that possess certain well-marked differentiating characters, and which propagate true from seed.
Race, altitudinal, a race adapted by inheritance to a certain altitudinal belt.
Race, climatic, a race adapted by heredity to specific climatic conditions.
Race, geographic, a race peculiar to a definite geographic region.
Race, local, an aggregation of individuals of a given species by inheritance adapted to a given environment more perfectly than other groups of the same species.
Race, soil, a race peculiarly adapted by inheritance to a specific kind of soil.
Raceme, a flower cluster, with stalked flowers arranged along the sides of an elongated axis (as in prunus serotina).
Rachis, the axis of a compound leaf, or of a spike or other indefinite inflorescence.
Racemose, in racemes, or resembling a raceme.
Rachilla, the axis of the spikelet in grasses.
Radiant, with the marginal flowers enlarged and ray-like.
Radiate, with ray-flowers; radiating.
Radical, belonging to or arising from a root, or from a root-like portion of the stem below the ground.
Radicle, the portion of the embryo from which the root develops.
Radices adligantes, clinging roots.
Radices columnares, columnar roots.

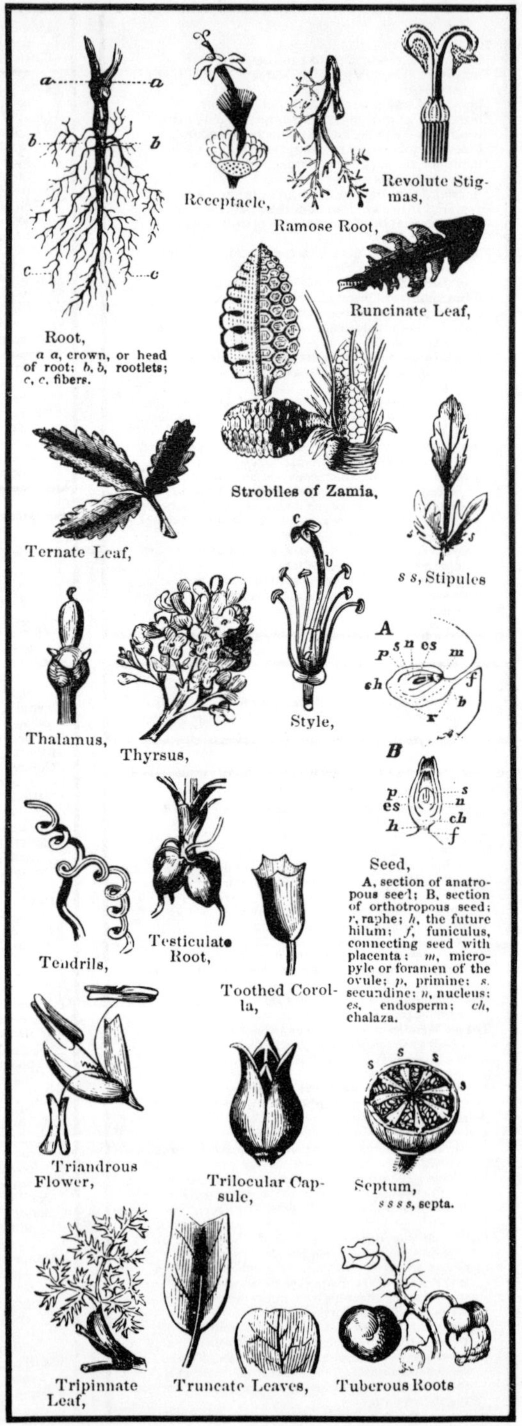
a a
b b
c c
Root,
a a, crown, or head of root: b, b, rootlets; c, c, fibers.
Receptacle,
Ramose Root,
Revolute Stigmas,
Runcinate Leaf,
Strobiles of Zamia,
Ternate Leaf,
s s, Stipules
Thalamus,
Thyrsus,
Style,
A
p s n es m
ch
f
b
r
B
p
es
h
s
n
ch
f
Seed,
A, section of anatropous seed; B, section of orthotropous seed; r, raphe; h, the future hilum; f, funiculus, connecting seed with placenta: m, micropyle or foramen of the ovule; p, primine; s, secundine; n, nucleus; es, endosperm; ch, chalaza.
Tendrils,
Testiculate Root,
Toothed Corolla,
Triandrous Flower,
Trilocular Capsule,
Septum,
s s s s, septa.
Tripinnate Leaf,
Truncate Leaves,
Tuberous Roots

GLOSSARY

Radices fulcrantes, stilt-like roots.
Radices parietiformes, or tabular roots.
Radices tuberosae, or tuberous roots; roots beset with tuber-like enlargements.
Radix, the root.
Raphe, that part of the stalk of an anatropous ovule which is fused with the body of the ovule; in diatoms, the median line on the frustule, possibly a slit.
Ray, one of the peduncles or branches of an umbel; the flat marginal flowers in compositae.
Receptacle, of a flower; the abbreviated or flattened axis upon which the various floral members are inserted.
Resilient, springing back, rebounding: used of fruit-stalks, stamens, etc.
Resin-duct, an intercellular passage into which resin is secreted and where it is stored.
Respiration, the term applied to the absorption by a plant of free oxygen from, and evolution of carbon dioxide into the air. It is the outward sign of a destructive oxidative process going on within the plant, by means of which latent energy is rendered available.
Reticulate, arranged as a network.
Retrorse, turned backward or downward.
Retuse, with a shallow notch at the end.
Revert, reversion, a sudden return or breaking back to an ancestral form.
Rhizoids, the hair-like filaments of mosses and liver-worts, which perform the functions of roots.
Rhizome, an underground (or prostrate) stem of root-like appearance from which roots and herbaceous stems arise.
Rhizomorph, name given to the curious vegetative phase of agaricus melleus, which resembles a root.
Rhizophore, a leafless branch of peculiar construction which, in selaginella, arises at the place where ordinary branching takes place, and bears roots at its free end.
Rhizotomoi, a guild of herbalists in ancient Greece.
Ring, annual, the zone of wood formed from the cambium in the course of one season in a conifer or dicotyledon.
Ringent, gaping, as the mouth of a bilabiate corolla.
Ringing, the act of removing from a branch or trunk a circular zone of bark right down to the wood.
Root-cap, the cellular cushion produced at the apex or tip of a root.
Root-stock, same as rhizome.
Rostellum, the morphological apex of the gynoeceum of an orchid; usually a beak forming the boundary between the stamen and stigma in orchids.
Rosulate, collected in form of a rosette.
Rotate, wheel-shaped; circular and horizontally spreading.
Runner, a prostrate filiform branch which is disposed to root at the end or elsewhere.

Sac, a pouch, especially the cavities of anthers.
Saccate, with a pouch or sac.
Sagillate, like an arrow-head, with the lobes turned downward.
Saprophyte, a plant which grows on dead and decaying organic matter.
Samara, a nonsplitting winged or key fruit (as in acer, fraxinus, and ulmus).
Scabrous, rough to the touch.
Scale, a minute, rudimentary or vestigial leaf.
Scape, or scapus, a peduncle rising from the ground.
Scarification, the wearing down by abrasion of an outer more or less impervious seed coat, to facilitate water absorption and to hasten germination.
Scarious, thin, dry, and membranaceous, and not green.
Schizocarp, a polycarpellary fruit which breaks into 1-seeded portions.
Sclerotic-cell, a hard, thick-walled cell, often of irregular form; sclerotic cells may be united together into layers, or isolated in soft parenchyma.
Sclerotium, in fungi a tuber-like mass of hyphae, which, after remaining dormant for a while, ultimately sprouts, producing fructifications. In the myxomycetes it is the resting-stage of the plasmodium.
Scorpioid cyme, a definite inflorescence rolled up towards one side like a crook: common in boraginaceae.

Scutellum, the sucker or cotyledon of a grass embryo.
Scutiform, having the form of a shield.
Seed, the mature ovule, resulting from fertilization and consisting of the embryo, usually endosperm, and seed coats.
Seed certification, guarantee of seed character and quality by an officially recognized organization, usually evidenced by a certificate including such information as genuineness of species and variety, origin, purity, soundness, and germinative capacity.
Seed coat, the covering of the seed. Syn. integument. See testa.
Seed coat, impermeable, a seed coat which is resistant to the absorption of water, oxygen, or both.
Seed extraction, separation of seed from fruit by artificial means.
Seed origin, the locality in which the seed was collected. See provenience.
Seed pretreatment, any process, such as soaking, stratification, scarification or acid treatment, to which seeds are subjected to improve and hasten germination.
Seed setting, formation of a seed crop on plants.
Seedling, a young plant grown from seed. In a nursery, a young tree which has not been transplanted.
Semesan, an organic mercury compound for treating seeds. See uspulun.
Semifrutex, or semi-shrub, a shrub the shoots of which become woody at the base only, this portion alone being perennial.
Sepal, a leaf-member of the calyx.
Sepaloid, resembling a sepal.
Separation-layer, see absciss-layer.
Septum, a partition; a thin wall separating compartments.
Sericeus, silky; clothed with soft straight hairs.
Serotinous, flowering or fruiting late in the season, as in autumn. Also, bearing persistent concs or fruits.
Serrate, of leaf-margins; beset with teeth pointing towards the apex.
Serrulate, diminutive of serrate; serrate with small teeth.
Sessile, without a stem or stalk.
Seta, a bristle; the stalk of the spore-capsule in a moss or liverwort.
Shell, the hard layer of a nut as distinguished from its hull and kernel.
Shoot, that portion f the plant which is differentiated into stem and leaves and bears the reproductive organs.
Sieve-cells, cells which have pores in their walls causing a sieve-like appearance; sieve-tubes.
Sieve-plates, areas in the walls of sieve-cells or sieve-tubes perforated by pores.
Sieve-tube, an articulated tube whose contiguous elements communicate by means of open pores aggregated together upon sieve-plates. The sieve-tube is the characteristic element of the phloem.
Siliqua, the fruit of a cruciferous plant, a longish pod or seed-vessel.
Sinistrorse, used of twining stems which turn from north through west to south, etc.: the opposite of dextrorse.
Sinuous, sinuate, used of a leaf-margin which is strongly indented in a wavy manner.
Sinus, the space between the lobes of a leaf.
Sling-fruit, a general term given to any fruit which, in virtue of the possession of contractile tissues, throws its seeds to a distance, or is itself so thrown.
Soboles, a thin creeping stem, often subterranean.
Soredium, the 'brood-body' or 'brood-bud' of a lichen, consisting of a few algal cells wrapt round with a weft of fungal hyphae.
Sorus, a cluster of sporangia, such as those of ferns.
Soundness, percent of seeds which are fully developed, or sound.
Spadiciform, like a spadix.
Spadix, a fleshy spike.
Spathe, a large bract-like sheath inclosing an inflorescence.
Spatulate, like a spatula, oblong with the lower end attenuated.
Species, a group of individuals with so many characteristics in common as to indicate a very high degree of relationship and a common descent; the unit of plant (or animal) classification and designated by a binomial Latin name.
Spermatium, a male sexual cell which becomes free, but is unprovided with special organs of locomotion.
Spermatoplasm, the protoplasm of the male sexual cell.
Spermatoplast, a male sexual cell.
Spermatozoid, a free-swimming male sexual cell provided with cilia as organs of locomotion.

Spike, an elongated cluster of stalkless flowers (as in amorpha).
Spikelet, diminutive of spike; especially applied to flower-clusters of grasses and sedges.
Spine, a sharp-pointed body possessing vascular tissue, commonly a branch of some portion of a leaf.
Spinose, with spines or similar to spines.
Spinule, a small sharp projection.
Spinulose, with small sharp processes or spines.
Sporange, a sac containing spores.
Sporangiole, in the fungi, a small sporangium usually containing few spores, and larger many-spored sporangia being also present.
Sporangiophore, that which bears sporangia; a scale bearing sporangia in equisetum.
Sporangium, a sac within which spores are developed.
Spore, a reproducitve cell which becomes free, and is capable of developing into a new individual.
Sporidium, a spore abjointed from a pro-mycelium.
Sporocarp, a fructification, often the result of a sexual act, in which spores are produced, as in Red Sea-weeds and fungi. Also used of the sporangial receptacles of the hydropteridae.
Sporogonium, in mosses; the so-called moss-fruit with its appendages, consisting mainly of the capsule and seta or stalk.
Sporophyte, that stage in the life-cycle of a plant which bears the spores. Cf. oophyte.
Spur, an excavated slender continuation of some portion of a flower, usually containing nectar.
Squamiform, scale-like.
Squamigerous, furnished with scales.
Stamen, the pollen-bearing or male organ of a flower.
Staminate, male; bearing stamens or pollen-producing organs but no pistils.
Staminiferous, bearing stamens.
Staminode, a sterile stamen.
Standard, in papilionaceous flowers, is the unpaired, posterior petal.
Sterigma, the tube or stalk-like branch from which conidia are abstricted.
Sterile, without spores, or without seed.
Stigma, the part (usually the tip and mostly sticky or hairy) of the pistil which receives the pollen.
Stipe, the stalk of an organ.
Stipilate, provided with a stipe.
Stipulate, with stipules.
Stipules, paired foliaceous appendages of the leaf-base.
Stirps cirrhosa, a tendril-bearing stem.
stirps clathrans, a lattice-forming stem.
Stirps fluctuans, a floating stem.
Stirps humifusa, a prostrate stem.
Stirps palaris, a standard-stem, i.e. an erect, unbranched stem.
Stirps plectens, a weaving stem.
Stirps radicans, a stem which climbs by means of roots.
Stirps volubilis, a twining stem.
Stock, the parent forms from which a hybrid is derived.
Stock, class of, age of nursery stock denoted by two or more figures (as 2-0, 2-1, or 1-1-1), the first figure indicating years in the seedbed and the succeding figure or figures the years in the transplant bed or beds.
Stolon, a trailing or reclining stem above ground, which strikes root where it touches the soil, there sending up new shoots which later may become separate plants.
Stoloniferous, bearing stolons.
Stoma, an intercellular space or pore in the epidermis which, bounded by adjustible guard-cells, forms the means of communication between the lacunae of the plant and the outside air.
Stone, a hard, bony fruit part containing the seed (as in cornus and prunus).
Strain, a group of cultivated plants differing from the race to which it belongs by no apparent morphological characters, but by some enhanced, or improved physiological tendency propagated from seed.
Stratification, the operation or method of burying seeds, often in alternate layers, in a moist medium, such as sand or peat, to overcome dormancy.
Stratification, cold, stratification at low temperatures, generally just above freezing, usually to overcome embryo dormancy.
Stratification, warm, stratification at higher temperatures, usually about room temperature, chiefly to overcome seed-coat dormancy.
Striate, marked with fine, longitudinal lines or ridges.

Strict, Straight and erect.
Strigose, with appressed or ascending stiff hairs.
Strobile, a multiple fruit in the form of a small cone or head (as in alnus and betula). Also, a cone.
Stroma-starch, in certain algae (e.g. hydrodictyon), the fine-grained starch deposited throughout the chlorophyll-body, which plays a different part in the economy of the plant from that deposited around the pyrenoid.
Strophiole, an appendage to a seed at the hilum.
Strophtolate, with a strophiole.
Style, the stalk of the pistil between the ovary and stigma.
Stylopodium, the expanded base of a style.
Subacute, somewhat acute.
Suberin, a corky substance; the substance or group of substances present in cuticularized or corky cell walls.
Subex, a stem bearing scale-leaves.
Subfalcate, somewhat scythe-shaped.
Subligneous, somewhat woody in texture.
Subterete, nearly terete.
Subulate, awl-shaped.
Subversatile, partly or imperfectly versatile.
Succulent, fleshy, pulpy.
Sucker, a branch or shoot from a creeping underground stem or root which ascends above ground and tends eventually to become a separate individual plant.
Suffrutex, an under-shrub; a woody plant of quite humble growth.
Suffruticose, somewhat shrubby.
Sulcate, grooved longitudinally.
Superior, applied to the ovary when free from the calyx; or to a calyx adnate to an ovary.
Surculus, or Sucker, a shoot arising from a subterranean base.
Suspensor, in flowering plants and in selaginella; the filament of cells at the lower extremity of which the embryo arises.
Suture, a line of splitting.
Swarm, a social aggregate of simple organisms which live together but are not attached to any substratum.
Swarm-spore, a motile, ciliated, asexual reproductive cell destitute of a cell-membrane.
Switch-plant, a plant with reduced or wanting leaves, the shoots of which are green and subserve the functions of leaves.
Symbiosis, the association of two organisms which live together in intimate connection, both contributing to their mutual welfare.
Syncarp, a fleshy multiple fruit, consisting of the enlarged ovaries of several flowers more or less united (as in morus).
Syncarpous, said when the carpels of a gynoeceum are united.
Synconium, the fleshy excavated inflorescence of a fig.
Synergidae, two naked cells situated at the micropylar end of the embryo-sac, and assisting in the passage of the male cell to the egg in progamic fertilization.
Syngenesia, the 19th class of the Linnean system.
Syngenesious, having coherent anthers.
Systole, the rhythmic contraction of a contractile vacuole.

Temperature, room, in laboratory work, 68 to 70 degrees F.
Tenaculum, the clasping, rosette-like clamps of struvea, by means of which independent branches are held together.
Tendril, a filamentous branched or unbranched organ, usually sensitive to contact, by means of which a plant climbs.
Tentacle, an irritable hair or emergence on a leaf, as in dionaea, drosera, etc.
Terete, round, i.e. circular in transverse section.
Ternary hybrid, the plant resulting from crossing a hybrid with a species different from either of its parent forms.
Ternate, used of compound leaves with three leaflets, one terminal and two lateral.
Testa, the outer, and usually the harder, seed coat. See seed coat.
Tetrad, a group of four cells (e.g. spores, pollen-grains), usually arranged in the four corners of a 4-sided pyramid (tetrahedon).
Tetradynamia, the 15th class of the Linnean system.
Tetradynamous, used of stamens when there are six, of which four are longer than the other two—as in cruciferae.

GLOSSARY

Tetraspores, the asexual spores of Red Sea weeds, usually aggregated in clusters of four.
Thalamus, the floral receptacle.
Thallidium, a vegetative reproductive body, especially amongst thallophytes and muscineae.
Thallus, a vegetative body without differentiation into stem and leaf.
Tissue, a continuous aggregate of cells having a common origin.
Tomentose, felty or invested in tomentum.
Tomentum, dense matted investment of woolly hairs.
Torus, (1) the floral receptacle; (2) the thickening on the pit-closing membrane of a bordered pit.
Trabeculae, folds or ridges projecting into a cell from the wall; the term also given to strings of filamentous cells bridging intercellular spaces.
Tracheids, elongated, pointed, and more or less lignified cells occurring in wood.
Transpiration, the act of exhaling aqueous vapour from foliage or other portions of plants.
Tree percent, the percent of viable seed sown which produce usable (ordinarily 1-0) seedlings.
Torsion, twisting of an organ.
Tortuous, twisted or bent.
Tracheae, The canals or ducts in woody tissue.
Triandrous, with three stamens.
Tricarpous, composed of three carpels.
Trichoblasts, fusiform hard-walled cells.
Trichogyne, the filamentous portion of the female sexual apparatus of a Red Sea weed, which receives the spermatia.
Trichome, a hair-like or similar outgrowth of the epidermis.
Trimorphous, flowers with stamens of three different lengths or kinds; in three forms.
Triquetrous, three-sided, the sides channeled.
Truncate, appearing as if cut short at the tip.
Trunk, a main stem.
Tuber, a subterranean, somewhat fleshy shoot.
Tubercle, a small excrescence.
Turgescence, Turgidity, the state of tension set up within a cell owing to the pressure of the osmotic cell-contents upon the elastic cell-wall.
Turion, a subterranean budding shoot, especially in perennials.

designated by the abbreviation "var." (varietas) preceding the Latin name. See Form.
Vascular bundle, a continuous strand of vascular tissue, consisting either of xylem or phloem, or of both. Not infrequently sclerenchymatous elements are associated with the bundle, when it is termed a fibro-vascular bundle.
Vascular elements, cells or vessels whose main function is the distribution of water or formed food-substances. The chief of them are the vessels and tracheids of the wood, and the sieve-tubes of the phloem.
Vein, one of the branches of the woody portion of leaves or other organs.
Veinlet, a branch of a vein.
Velum, in isoetes; the indusium-like membrane which covers the sporangium.
Velum partiale, in hymenomycetes; the veil stretching from the stipe to the edge of the pileus. It often remains as the annulus.
Velum universale, in hymenomycetes; the membranous wrapper inclosing the whole fructification.
Venation, the arrangement or pattern of the vascular bundles in a leaf.
Veneer grafting, grafting by beveling the scion and fastening it to a groove on the stock.
Ventral canal-cell, the small cell which is cut off from the central cell of an archegonium immediately below the neck.
Ventricose, unequally swollen.
Vernation, the arrangement of the parts in the bud, especially a vegetative bud.
Verrucose, covered with warts.
Versatile, turning freely on its support
Verticillate, arranged in a whorl.
Vessel, a tube consisting of cells which have become confluent by the partial or complete absorption of the intervening walls. They are common in the wood of angiosperms.
Viability, the potential capacity to germinate.
Viviparous, a term applied to plants the seeds of which germinate whilst still on the parent plant.
Volva, same as velum universale.

Umbel, an umbrellalike flower cluster with the pedicels arising from the same point (as in aralia nudicaulis).
Unguiculate, narrowed at the base into a claw: used of petals.
Urceolate, hollow and contracted at or below the mouth like an urn.
Uspulun, an organic mercury compound (hydroxymer-curichlorophenol sulfate and similar compounds) used for disinfecting seeds and seedbeds. Ceresan and semesan are somewhat similar preparations.

Whorl, an arrangement of three or more organs (such as leaves or branches) in a circle around the axis.
Wing, a membranous or thin and dry expansion or appendage of a seed or fruit.
Winnow, to separate and drive off chaff and debris as by fanning.
Witches' Broom, a form of gall found on the silver fir and other conifers; sometimes applied to the bird's-nest-like hypertrophies on the birch, etc.
Wood, the hard, lignified portion of the vascular tissue otherwise known as the xylem. It contains tracheids, woody fibres, and wood parenchyma, though not all of these are necessarily found in the wood of any given plant.

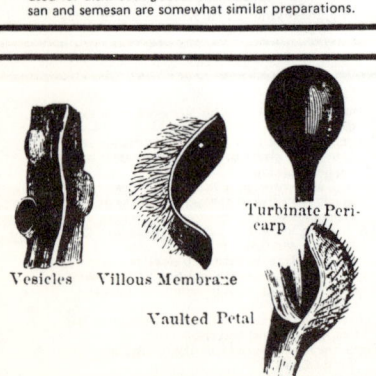

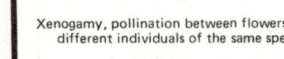

Xenogamy, pollination between flowers growing on different individuals of the same species.

Xenogamy, pollination between flowers growing on different individuals of the same species.
Xylem, the woody portion of vascular tissue. See wood.

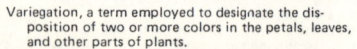

Variegation, a term employed to designate the disposition of two or more colors in the petals, leaves, and other parts of plants.
Variety, a subdivision of a botanical species having some characters different from the typical and

Zoogloea, a solid gelatinous colony of bacterial organisms.
Zygomorphic, applied to flowers which are symmetrical about one plane only, or can be cut into similar halves in only one plane.
Zygospore, a spore formed by the union of two gametes.
Zygote, a general term for the product of fusion of two gametes.
Zygozoospore, the motile stage of a zygote, the product of fusion of two motile gametes.